Contents

1. Introduction
2. Article 343. Hurricane Florence: cover-up for tidal force from object in outer space
3. Article 349. Planet X: gravitational and electric effects on the Planets
4. Article 364. Planet X causing the earth's surface to break up
5. Article 370. Hurricane Michael used to cover up tidal event due to Planet X approach
6. Article 371. Hurricane Michael: tidal event due to object approaching from space
7. Article 377. Pink clouds over Florida indicate something is very wrong with the Earth and the Sun

Books previously published

Book 1: Planet X: the awakening is now.
Book 2: The Planet X Report 2017: Photographic Evidence.
Book 3: Planet X Revealed Gravity and Light.
Book 4: The Sun Simulator
Book 5: Chemtrails: The Silent Killer.
Book 6: Planet X Physicist Articles: Part 1
Book 7: Planet X: The effects on the Earth and the Sun
Book 8: Planet X and the Solar System

Chapter 1
Introduction

This is a short book detailing the events around the landfall of two hurricanes which indicates that these events or storms are actually caused by the Planet X System objects, also called Stellar Cores, which seem to be continually arriving at the Earth and affecting it in several ways. The Planet X created storms are being covered up with artificially engineered storms, which nevertheless never reach hurricane strength. This indicates that the technology exists to create rainfall and lots of it; but producing full cyclonic circulation and extremely high winds, in the earth's atmosphere, is not possible. Full cyclonic circulation and very high winds only occur whenever an external gravitational force enters the earth's atmosphere, so these can only be produced by actual massive objects entering the earth's atmosphere. This is a very closely guarded secret by the powers that be which is exposed in this book.

Dr. Claudia Albers

Planet X physicist

October 15th 2018

Chapter 2

343. Hurricane Florence: a cover-up for the tidal force from an object in outer space

Hurricane Florence started impacting the East Coast of the United States on Thursday, September 13th, 2018. I am writing this article on September 14th, a day later. For many days before the Hurricane hit the coastline, there were many dire warnings about the strength of this hurricane. On Monday afternoon, September 10th, and thus 3 days before the hurricane reaching the coastline, it was predicted that it would reach the coastline as a category 3 hurricane. However, just before it did reach the coastline Florence was categorized as a category 1 hurricane.

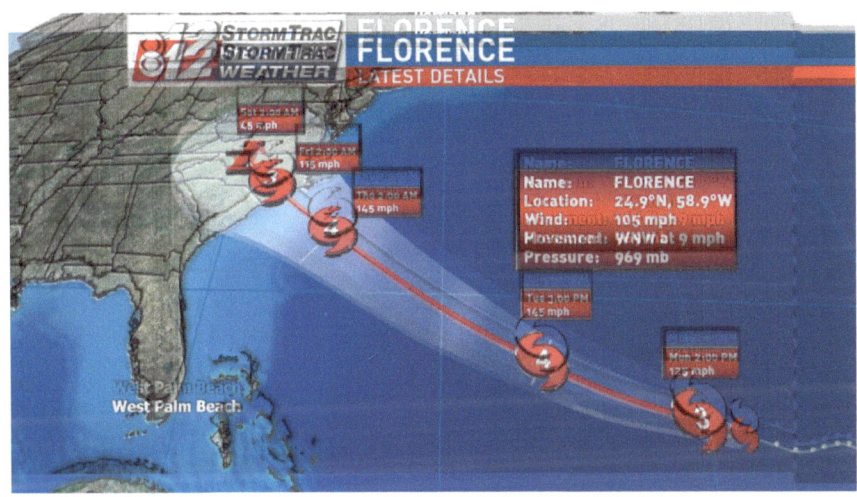

Figure 2.1: Florence predicted to reach the coastline as a category 3 hurricane on Monday 10th September 2018 [1].

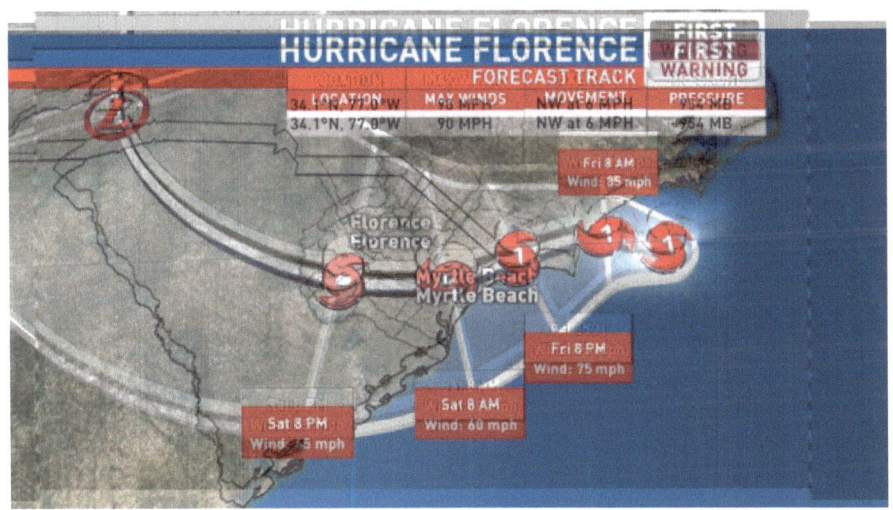

Figure 2.2: Hurricane Florence prediction on September 14th 2018 showing it hit the coast as a category 1 hurricane [1].

With the hurricane, category predictions come rainfall predictions, as well as, storm surge predictions. Storm surge refers to the predicted height of the water coming on-shore, from the ocean. This is attributed to the wave heights brought on by the wind generated by the hurricane. The forecast below is from September 11th, 2 days before the hurricane reached the shore, when it was still believed that the hurricane would come onshore as a category 3 hurricane.

Figure 2.3. Storm surge prediction from September 11th 2018, when it was believed the hurricane would reach shore as a category 3 hurricane. The predicted storm surge for Myrtle Beach was between 4 and 6 feet and for Wilmington, where the eye of the hurricane came ashore it was 6 to 8 feet, with 8 to 10 feet in a few places [1].

Figure 2.4. Storm surge prediction on the morning of September 14th continues to show that the predicted storm surge for Myrtle Beach is 4 to 6 feet, and for Wilmington, it is 7 to 11 feet, higher than 3 days before, even though it is now known that this is a category 1 hurricane [2].

In fact, the weather channel goes as far as to say on Friday morning that a storm surge of 7 to 11 feet has been verified.

Figure 2.5. Weather channel stating that a storm surge of 7 to 11 feet has been verified at New Bern [2].

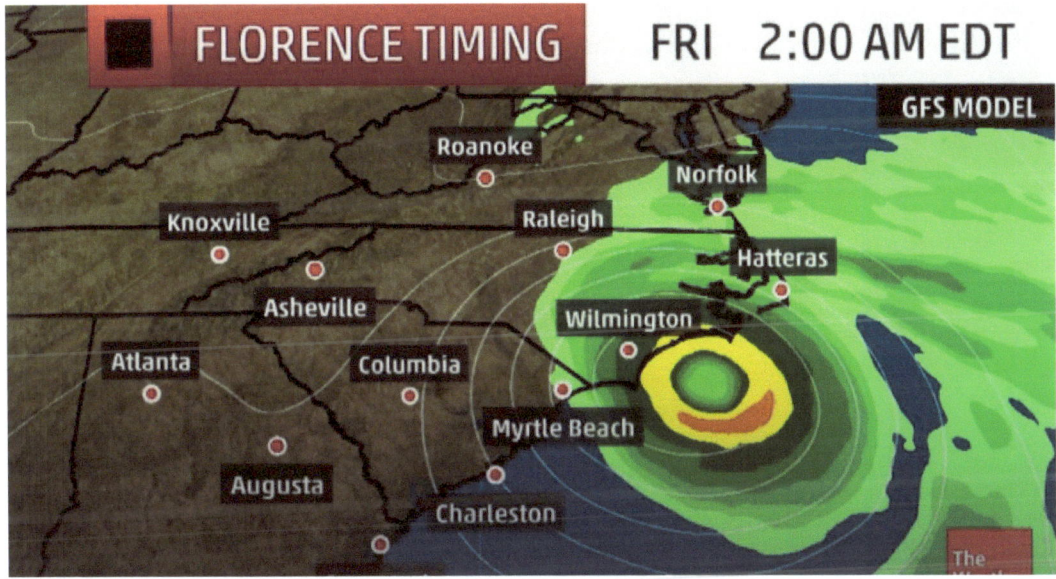

Figure 2.6. A screenshot from the weather channel showing that Hurricane Florence came ashore exactly where the highest predicted storm surge occurred [2].

But how is this possible? If storm surge is as a result of the waves due to the winds generated by a hurricane, and since the wind generated by a category 1 hurricane are much weaker than those generated by a category 3 hurricane, why have not the storm surge predictions been revised down at the same time that the category was? But instead it seems that the storm surge prediction was accurate, only the category of the hurricane was not.

Now, if a storm surge is related to the wind generated by a hurricane, then a drop in the wind would necessarily lead to a proportional drop in the storm surge. But, if the wind decreases in strength, and the storm surge does not decrease in strength, this means that the two variables are not related, in other words, one does not influence the other. Thus, the storm surge, in this case, could not have been created by the hurricane force winds but by something else. This suggests that this was, in fact, a tidal event covered up by a hurricane. As I have shown in Article 335: Biological and Ecological weapons in use against us [3], the weather has been controlled worldwide, for a very long time, and hurricanes can both be artificially created, as well as, steered, through the use of electromagnetic waves. Hurricane Florence, therefore, seems to have been created and steered with the express purpose to cover-up an exceptionally high tide event. The fact that the tidal surge was accurately predicted whilst the hurricane category, and therefore the wind generated by the hurricane, was not, indicates that the strength of this hurricane was inflated from the beginning, possibly indicating that artificially created hurricanes cannot reach the strength of naturally created hurricanes. If the tidal surge could be accurately predicted, it must have been possible to predict the strength of the hurricane more accurately as well, especially since it was artificially generated, even though the two events were only related by one being designed to cover up the other.

Now tidal events are not new, there have been several ocean recession events reported during 2017 and some this year. The first which seemed to have occurred at the Uruguay and Brazil coastlines could not have been associated with any storms, but it became clear that there was an attempt to cover up some of the occurrences after that, with storms.

Figure 2.7: Left: Ocean recedes leaving boats sitting on mud, in the harbor in Punta del Este, Uruguay, on August 11th, 2017. The ocean came back but this extreme low tide had never happened before. Right: An empty beach, due to the ocean receding, from the Brazilian coast, on August 12th, 2017, no large storms or hurricane could be blamed for the phenomenon, as there were no storms or hurricanes anywhere near this coastline. This too was unprecedented (see Article 227: Stellar Cores affecting the earth and possible connection to Volcanic Eruptions) [4].

At about the same time that the ocean was receding on the east coast of South America, there were huge waves crashing onto the Chilean coastline, indicating that both unnaturally high and unnaturally low tides were occurring on the planet, at about the same and were, therefore, most likely related.

Figure 2.8: Large waves crash into the Chile coastline. This was likely to be due to an abnormally high tide event.

Figure 2.9: The ocean receded in Santa Elana, Ecuador, on January 3rd, 2018, 6 days before a strong earthquake in Honduras, along a line joining two tidal events. A strong earthquake occurred in Honduras on January 9th, 2018, initially reported at a magnitude 7.8, but was later downgraded to 7.6, suggesting an association between tidal events and strong earthquakes (see Article 127: Honduras 7.8 magnitude earthquake on January 9th and water recession) [5].

Figure 2.10. People walk out onto the sand, in Tampa Bay Florida, on September 10[th,] 2017. The beach was left empty due to the ocean receding. A hurricane affecting the area passed overhead, before the water started returning, the next morning, showing that the ocean recession did not occur as a result of hurricane winds, offshore. It does, however, indicate that an attempt was made to cover up this tidal event with a hurricane (see Article 227: Stellar Cores affecting the earth and possible connection to Volcanic Eruptions) [4].

Hence, tidal events in the form of unnaturally low and unnaturally high tides have been observed and it is clear that at least in one instance there seems to have been an effort to cover it up with a hurricane, which we know can be artificially generated and steered. Why would such a tidal event be covered up? It would be covered up because tidal events can only be created by an object closely approaching the planet from outer space. A tide is produced by a tidal force, or a weak differential gravitational force, which can only be created by an object, which generates a gravitational field, closely approaching earth.

The moon creates earth's normal tides as illustrated below. Abnormal tides, in other words, tides higher than normal moon generated tides, can only be created by an object, other than the moon, approaching our planet. This object may then also cause strong earthquakes and induce volcanic eruptions, as a result of its gravitational influence.

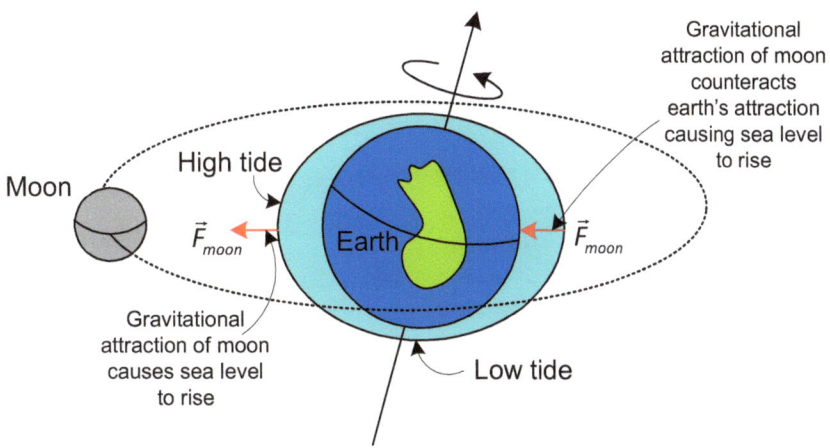

Figure 2.11. The moon's effect on ocean tides on earth: The moon's influence is tidal in nature: the attraction on the ocean, directly beneath the moon, is much stronger, than on the ocean, further away. This causes a difference in the gravitational attraction over adjacent regions.

Thus, a tidal event, as a result of an object approaching earth, is most likely in progress right now and is the cause of what is being termed storm surge. This storm surge is, in fact, a tidal surge, due to an object in space. This tidal surge is likely to continue for the next couple of days as the object continues to influence the area, with its gravitational field, which is most likely why hurricane Florence has been made to move so slowly. This means that the object is not likely to move far from that one spot over the surface of the earth. An object which is able to stay stationary over a region on a planet can only be a Stellar Core, as these objects have been observed to remain over one spot, in the Sun's corona, for long periods of time, and Stellar Cores observed in the sky, from earth, have also remained in the same position for extended periods of time.

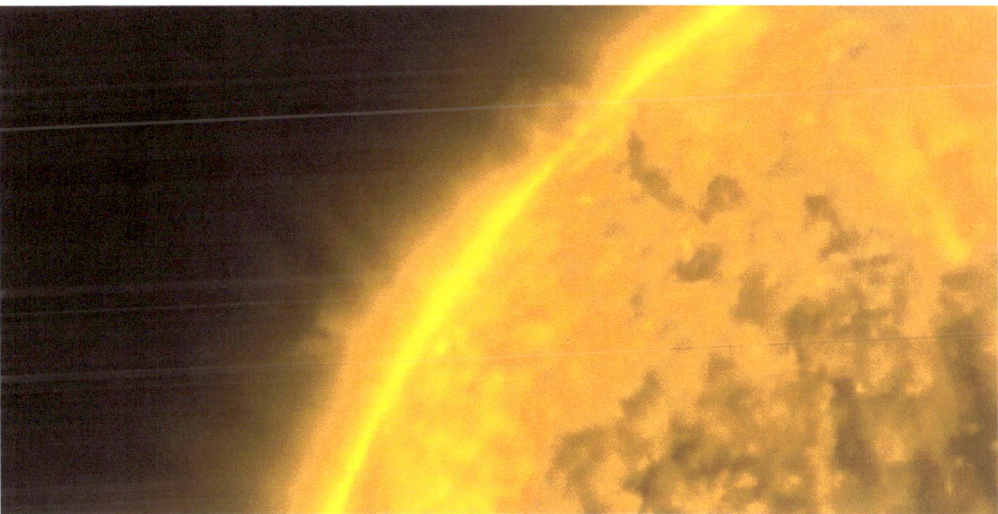

Figure 2.12: SDO image in the 171 angstrom wavelength from October 13th, 2017 showing a dark Stellar Core, which appears to be about half of the size of Jupiter making a matter connection with the Sun.

Figure 2.13: An object which seems to be emitting red light, and is surrounded by a diffuse cloud, is seen here in a European webcam. It was caught by Jeff P in early March 2018. The object did not move across the sky but remained in the same position for an extended period of time. Stellar Cores are also often observed to stay stationary with respect to a point on the Sun with which they have made a matter connection with (see Article 243: Earth hosting at least 3 Planet X System Objects) [6].

It is therefore likely that one of these objects has approached Earth at this time and is remaining close to the area where artificially engineered Hurricane Florence came ashore.

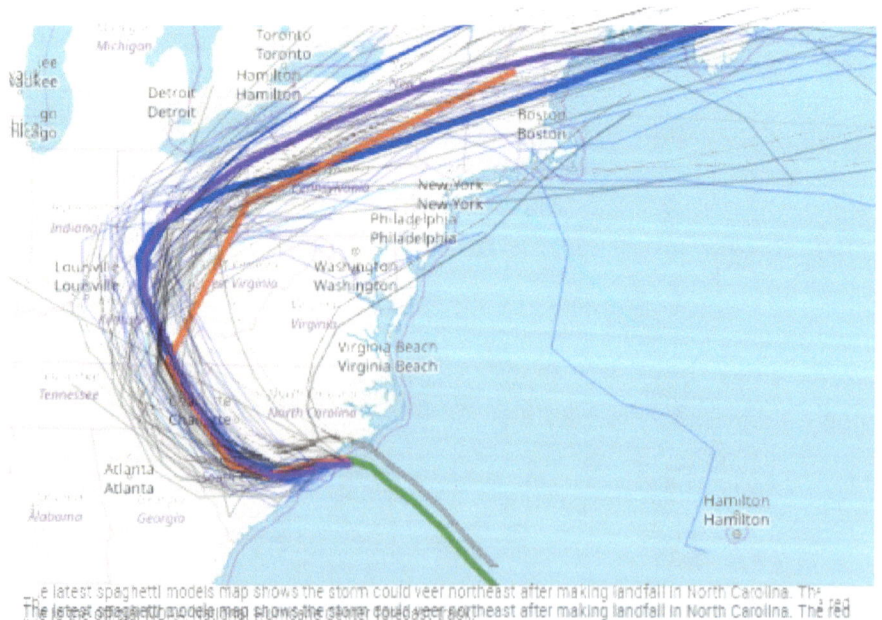

Figure 2.14: Hurricane Florence's predicted path over the next few days.

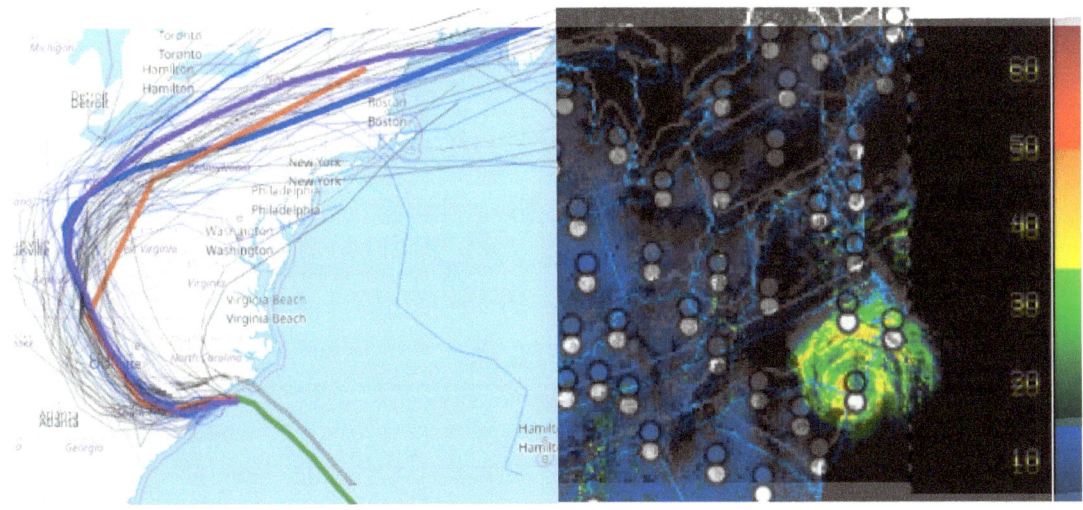

Figure 2.15: Predicted path of Hurricane Florence (left) and the Nexrad radar stations (right) that may be used to energize it, so that it doesn't dissipate, or fall apart, as well as, steer it, with the aim of generating inland flooding.

In conclusion, Hurricane Florence most likely never reached the strength that it was reported to have, from the time that it was artificially formed in the Atlantic. The Hurricane was most likely steered so that its arrival coincided with the influence of an object, in space, close to earth, exerting a gravitational force on Earth's oceans, and thus creating a tidal event, which impacted the east coast of the US where Hurricane Florence came ashore. The object must be a Stellar Core and thus a member of the Planet X System of dead stars.

References:

[1]   https://wpde.com/news/local/gordon-forms-florence-heads-west

[2]   https://weather.com/storms/hurricane/video/storm-surge-and-flooding-threat-from-hurricane-florence

[3]   Albers, C. (2018). Article 335: Biological and Ecological weapons in use against us (in Book 8: Planet X and the Solar System).

[4]   Albers, C. (2018). Article 227: Stellar Cores affecting the earth and possible connection to Volcanic Eruptions (in Book 6: Planet X Physicist Articles Part 1).

[5]   Albers, C. (2018). Article 127: Honduras 7.8 magnitude earthquake on January 9th and water recession.

[6]   Albers, C. (2018). Article 243: Earth hosting at least 3 Planet X System Objects (in Book 7: Planet X: The effects on the Earth and the Sun).

Chapter 3

349. Planet X: gravitational and electric effects on the Planets

I have written many articles detailing the evidence for the existence of objects in the Sun's corona. These objects are not like Solar System objects, they are dead stars, planets, and moons. They are dead because they are energy depleted. From observing these objects, I discovered that gravity is quite different from what I had been taught; gravity is dependent on the photon energy, inside the particles of a celestial body, so that when an object is an energy depleted, its ability to interact gravitationally is very low (see Article 210: Stellar Core gravity: tidal and G is not constant) [1]. This means that these objects lose the ability to hold on to their outer layers of material, and so shed them, until their solid dense cores are exposed (see Article 320: Planet X debris incoming: sprites and earthquakes) [2]. This is why I have called these objects Stellar Cores, although only the large ones were once living stars, the smaller objects are actually planetary and moon cores (see Article 319: Planet X System: planets, moons, and debris) [3].

Figure 3.1. On the left: SDO image in the 171 angstrom wavelength from October 15th, 2017. Center and right images show the same object in false color. The Stellar Core in the Sun's corona was caught by Scott C'one.

As the objects are low in energy, they absorb energy from the Sun, and also absorb material from the Sun, this material then envelops the object and emits light due to the energy absorption process, or what I call the Planet X effect, which leads to increased temperature and ionization of the material (see Article 338: The Planet X effect: heating and ionization in contact regions) [4].

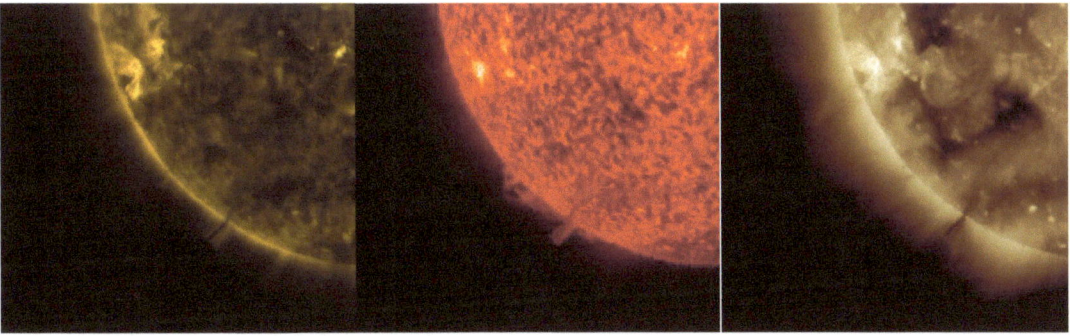

Figure 3.2. Images of the Sun, as detected by the SDO satellite, on March 11th, 2012, at 6:34 (UTC), in the 17.1, 30.4 and 19.3 nm (ultraviolet) wavelengths. A dark spherical object is seen drawing material through a gravitational vortex connection with the Sun. The object is about half the radius of Jupiter (5 times larger than the Earth). The dark root like the connection is not as dark in the 304 angstrom image suggesting that the matter in it, comes from deep within the chromosphere, the Sun's liquid layer which therefore behaves like the ocean on earth (see Article 99: Planet X and Stellar Structure) [5].

These objects are also affecting the planets. As I have shown in Article 347: Gravity wave on Venus suggests Planet X presence [6], at least one of these objects is being hosted by Venus and is the cause of the stationary gravity wave, which was first observed in Venus' atmosphere, in 2015. The earth is also hosting several of these objects, at least 3 (see Article 243: Earth hosting at least 3 Planet X System objects) [7]. These 3 objects are likely to be small Stellar Cores, most likely moon cores and thus the size of small moons. But there may be a larger Stellar Core orbiting the Earth which may be much larger than the Earth. These objects will be absorbing energy from the earth leading to it losing gravitational energy and expanding as a result, causing its surface to the fissure.

Figure 3.3. Image obtained from a video by the Youtube channel Jeff P. The image comes from a web camera over Germany from October 31st, 2016. Three light sources can be seen in the image. The top one is orangey pink, the middle, and brightest, is white, edged by pink light, and the lower one is white. Chemtrail clouds in front of the objects show that they are real objects in the sky (see Article 243: Earth

hosting at least 3 Planet X System objects) [7]. See Article 226b: Sun simulating devices: the irrefutable evidence [8] for details on Sun simulators.

What effects would these objects have on the earth? They are producing tidal events. Ocean recession and high tide events, as well as storms, earthquakes and volcanic eruptions:

Figure 3.4: The unprecedented recession of the ocean in Kholmsk, Sakhalin Island, Russia, on March 20th, 2018 (see Article 188: What is causing the ocean to recede all over the world?) [9]

Then, on March 21st, 2018, the Ebeko volcano, in the Kuril Islands, and thus just north of the region, where the ocean recession occurred, suddenly erupted, suggesting a connection between the two events:

Figure 3.5: Left: The Ebeko Vocano, on Paramushir Island, Kuril Islands, Russia, erupted on March 21st, 2018, a day after the water recession event. **Right:** Location of Ebeko Vocano in the Kuril Islands, Russia (see Article 227: Stellar Cores affecting the earth and possible connection to Volcanic Eruptions) [10]:

Tidal effects, which give rise to noticeable sea level changes, which have now been reported around the world, can only be produced by a massive object approaching earth, from space. But with every gravitational field, an electric field and therefore electrical effects will be induced. This is because the same charged particles, protons and electrons, give rise to both interactions, but in opposing ways. The gravitational interaction causes protons and electrons to repel, so they move apart, and set up an electric potential, and thus, an electric field. A high electric potential gives rise to electric discharges, which causes photons to leave particles, these photons then split into particles with mass. This is a matter creation event; the emerging massive particles then create a gravitational field. So gravity leads to the creation of an electric field and an electric field leads to the creation of a gravitational field. In other words, you cannot have a massive object with gravitational influence closely approaching the earth, without getting electrical effects, such as increased discharging (lightning) towards the earth and in the earth's atmosphere.

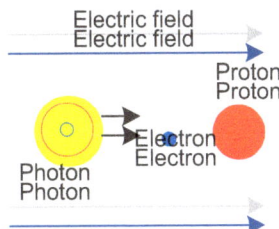

Figure 3.7: A photon moving through a region of electric field splits into its constituent particles

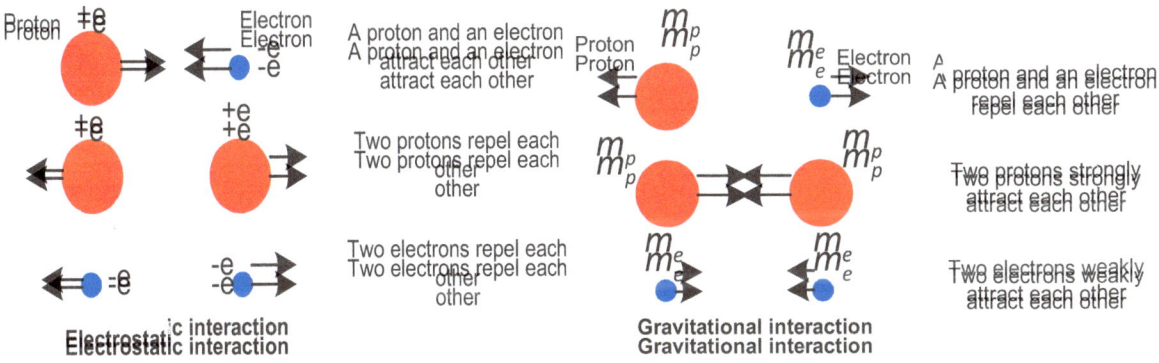

Figure 3.8: The electrostatic and gravitational interactions between protons and electrons. The electrostatic interaction is of equal strength in all 3 cases but the gravitational does not. The strength of the interaction is dependent on the energy of the photon and on the mass of the particles. It is this asymmetry which allows the universe to have the observed structure where all objects from atoms to galactic nuclei have a dense proton rich and positively charged interior and a negative electron outer layer (see Book: Planet X Reveals Gravity and Light for more details) [11].

You also cannot generate an electric field, without inducing gravitational effects. However, in order for an electric potential to give rise to a measurable gravitational field, much higher electric potentials than are normally used in our daily lives are needed. The effects generated by an astronomical object approaching the earth can be artificially induced, with the application of electromagnetic fields (see Article 335: Biological and Ecological weapons in use against us) [12], but these cannot have the strength and severity that naturally produced events resulting from the approach of a celestial object can have.

As I showed in Article 348: Venus bulge: gravitational waves and hollow planets [13] massive celestial objects, with cores, give rise to a gravitational field, which is in the form of a stationary gravitational wave, which produces both high and low sea levels, on the surface of the earth's oceans, when such a massive celestial object closely, approaches the surface of the Earth. The shape of this wave suggests that all cores, and thus all planets, moons, and stars are hollow in the center.

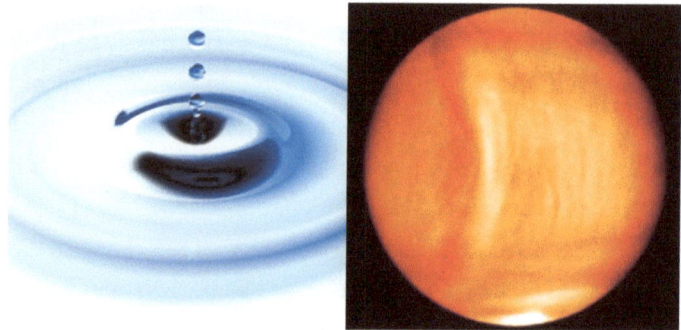

Figure 3.9. **Left:** A water wave with a central minimum of lower amplitude than the 1st maximum and lower amplitude than the 1st minimum, generated by a drop of water falling on a water surface. **Right**; The stationary gravitational wave on Venus: It seems to carry on to the back of the planet. The shape of this wave suggests that a gravitational wave will have a central minimum but its amplitude will be less than that of the first maximum and the first minimum.

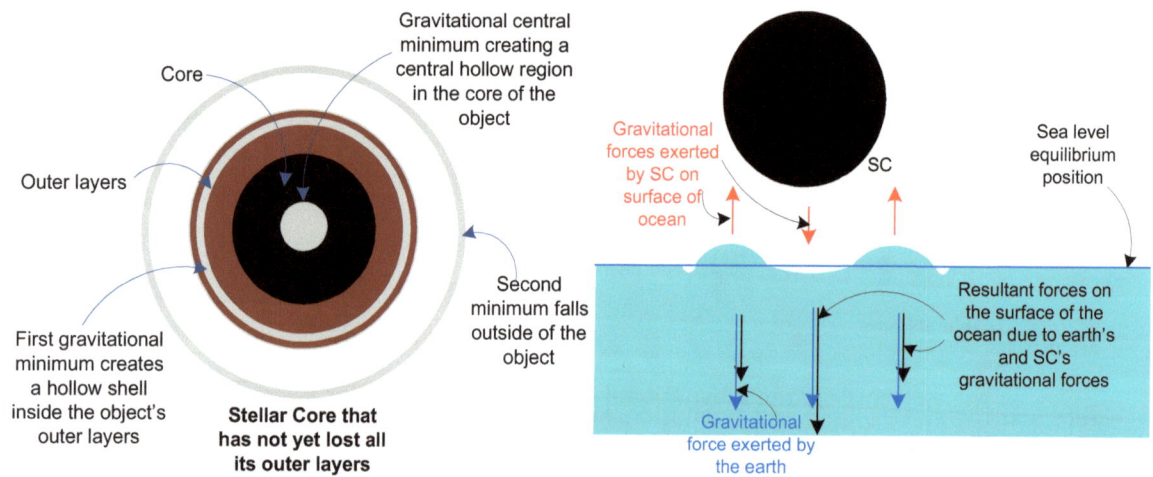

Figure 3.10. Left: A Stellar Core close to the surface of earth's oceans will produce a wave with a cross-section as shown. Right: All celestial objects with cores will have hollow centers, and shells, where the field reverses direction.

In conclusion, Planet X System Stellar Cores seem to have congregated mainly around the Sun, but at least a few seem to have congregated around Earth and Venus, and most likely, around all planets, in the Solar System. These objects seem to come very close to the Sun and the Earth, without causing cataclysmic events because they are energy depleted objects. However, they maintain enough gravitational energy to still have a gravitational effect, and as time goes on, that effect will increase as they gain energy, by absorbing it from Solar System objects. As the gravitational interaction does not occur without the electric, these effects are both gravitational and electric in nature.

References:

[1] Albers, C. (2018). Article 210: Stellar Core gravity: tidal and G is not constant (in Book 6: Planet X Physicist Articles: Part 1).

[2] Albers, C. (2018). Article 320: Planet X debris incoming: sprites and earthquakes.

[3] Albers, C. (2018). Article 319: Planet X System: planets, moons and debris.

[4] Albers, C. (2018). Article 338: The Planet X effect: heating and ionization in contact regions (in Book 8: Planet X and the Solar System).

[5] Albers, C. (2018). Article 99: Planet X and Stellar Structure.

[6] Albers, C. (2018). Article 347: Gravity wave on Venus suggests Planet X presence (in Book 8: Planet X and the Solar System).

[7] Albers, C. (2018). Article 243: Earth hosting at least 3 Planet X System objects (in Book 7: Planet X: The effects on the Earth and the Sun).

[8] Albers, C. (2018). Article 226b: Sun simulating devices: the irrefutable evidence (in Book 6: Planet X Physicist Articles: Part 1).

[9] Albers, C. (2018). Article 188: What is causing the ocean to recede all over the world? (in Book 3: Planet X Reveals Gravity and Light).

[10] Albers, C. (2018). Article 227: Stellar Cores affecting the earth and possible connection to Volcanic Eruptions (in Book 6: Planet X Physicist Articles: Part 1).

[11] Albers, C. and C'one, S. (2018). Book 3: Planet X Reveals Gravity and Light.

[12] Albers, C. (2018). Article 335: Biological and Ecological weapons in use against us (in Book 8: Planet X and the Solar System).

[13] Albers, C. (2018). Article 348: Venus bulge: gravitational waves and hollow planets (in Book 8: Planet X and the Solar System).

Chapter 4

364. Planet X causing the earth's surface to break up

In Article 343: Hurricane Florence: cover-up for tidal force from object in outer space [1], I showed that Hurricane Florence had been a cover up for the occurrence of a tidal event, at the East Coast of the United States, because the category of the hurricane had been sharply revised downwards and yet the forecasted storm surge had not. It was even worse than that, as NOAA used wind gust speeds, instead of sustained wind speeds to categorize the hurricane so that Hurricane Florence did not register any hurricane force winds, until it reached shore (see Article 345: Hurricane Florence tidal event and gravity waves) [2].

Figure 4.1: Weather channel stating that a storm surge of 7 to 11 feet has been verified at New Bern, on September 14th, 2018 [3]. This was higher than had been forecasted when it was still believed that the hurricane would reach shore as a category 4 hurricane. This was, therefore, a tidal surge, not a storm surge.

These tidal events seemed to have started in 2017, with ocean recession events, in Uruguay and Brazil. These were followed with several other similar events, in several parts of the world, including Venice, eastern Russia, Ecuador and Tampa Bay, Florida. The event in Tampa Bay Florida was also associated with the passing of a hurricane, which passed overhead before the water returned. At the same time, high tidal waves have suddenly on occasion appeared, such as in Durban South Africa, in March of 2017.

Figure 4.2: Left: Ocean recedes leaving boats sitting on mud, in the harbor in Punta del Este, Uruguay, on August 11th, 2017. The ocean came back but this extreme low tide had never happened before. **Right:** An empty beach, due to the ocean receding, from the Brazilian coast, on August 12th, 2017, no large storms or hurricane could be blamed for the phenomenon, as there were no storms or hurricanes anywhere near this coastline. This too was unprecedented (see Article 227: Stellar Cores affecting the earth and possible connection to Volcanic Eruptions) [4].

Figure 4.3: The ocean floods land in Durban South Africa in March of 2017. It was reported at the time that a hurricane of the coast of Madagascar was the cause but that is a ridiculous attempt at covering up a tidal event that can only be caused by an object coming from space.

Now, tidal events can only occur as a result of the gravitational influence of a celestial object, closely approaching earth (see Article 188: What is causing the ocean to recede all over the world?) [5]. Normal tides, on the earth, are mainly due to the moon's gravitational influence, so these abnormally high and abnormally low tides have to be as a result of massive objects closely approaching earth. The objects that would be likely to be responsible are, off course, the Stellar Cores belonging to the system, of the dead, or energy depleted, stars, and planets, which I have named Planet X System. These objects have been affecting the Sun and the earth for 100s of years. They enter the Sun's corona and the earth's atmosphere and create gravitational vortices. On earth, we call these gravitational vortices, hurricanes, tornadoes and water spouts. A tornado was first observed on earth in 1643 (see Article 361: Planet X producing gravitational vortices in the earth's atmosphere since 1643)[6].

Figure 4.4: SDO image in the 171 angstrom wavelength from October 13th, 2017 showing a dark Stellar Core, in the Sun's corona. A vortex connection appears below the object.

Figure 4.5. A tornado and a waterspout, a waterspout is basically a tornado over water. Notice the pink color, in the sky, indicating the presence of Stellar Core matter, which is emitting magenta (pink) colored light, in the left image. These gravitational vortices are the result of small Stellar Cores entering the earth's atmosphere just like they enter the Sun's atmosphere or corona.

Figure 4.6. Ocean recession event occurring at the same as high tide event, which was covered up with Hurricane Florence [2].

The ocean recession event, which occurred at the time of Hurricane Florence, led me to the understanding that the Stellar Cores, which were approaching the earth, and affecting it, gravitationally, were creating a wave pattern, on the surface of the ocean, with a region where the sea level was higher than normal, and another region, nearby, where the sea level was lower than normal. This meant that gravity was a stationary wave, which created regions of high and low density. Then, after observing a gravitational wave, which periodically appears in Venus' atmosphere, I understood that gravity is in the form of a wave, with a central minimum, in other words, the waves generated on the surface, of the ocean, would have a low sea level, central circular region, surrounded by a ring shaped high sea level region, followed by another low sea level ring shaped region (see Article 348: Venus bulge: gravitational waves and hollow planets) [7].

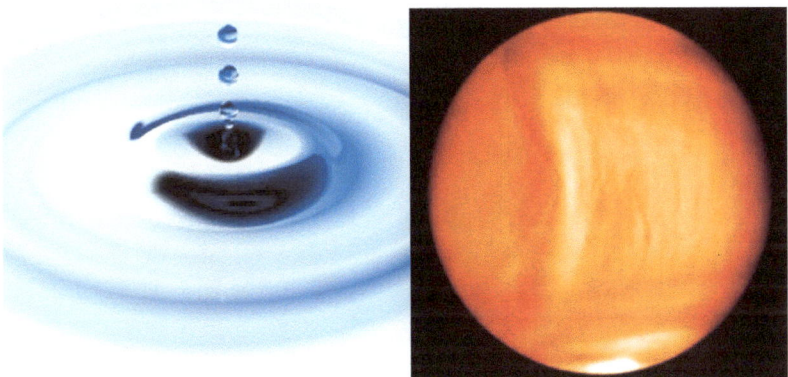

Figure 4.7. **Left:** A water wave with a central minimum of lower amplitude than the 1st maximum and lower amplitude than the 1st minimum, generated by a drop of water falling on a water surface. **Right**; The stationary gravitational wave on Venus: It seems to carry on to the back of the planet. The shape of this wave suggests that a gravitational wave will have a central minimum, but its amplitude will be less than that of the first maximum and the first minimum [7].

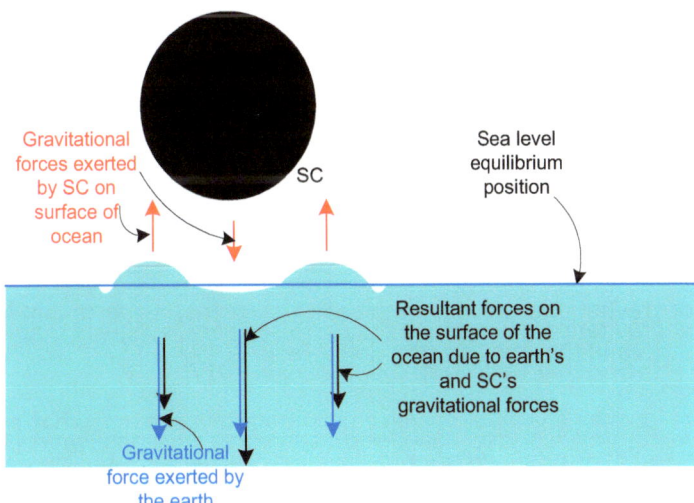

Figure 4.8. The stationary gravitational wave on Venus suggests that the gravitational wave pattern created by a Stellar Core, close to the earth's surface, has a central minimum [7].

So, these objects can create gravitational vortices, in the earth's atmosphere, and the objects that have been reaching the earth have now been creating gravitational waves, in the ocean, suggesting that these are larger, or gravitationally stronger objects, which can affect the earth from a greater distance. The gravitational vortex only forms, when the distance between the interacting massive objects is very small, as shown below; but the ring wave pattern would form from a much greater distance:

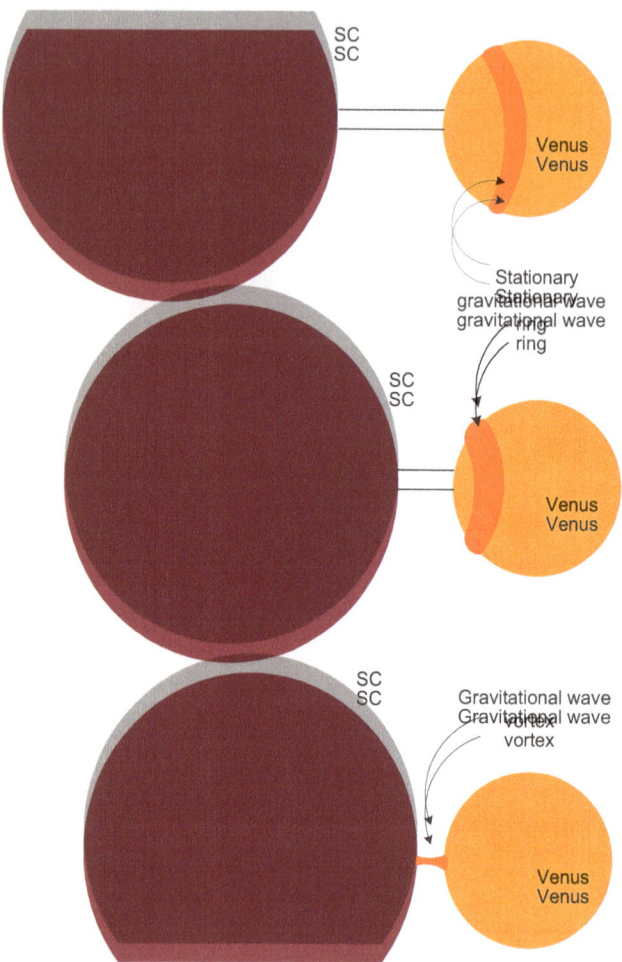

Figure 4.9: How a Stellar Core very close to Venus would produce a gravitational vortex, on its surface: The height of the wave increases, as the Stellar Core approaches the planet, at the same time that the radius of the ring decreases, until a vortex forms between the Stellar Core and the planet. This vortex is what is seen with the Stellar Cores in the Sun's corona because they are so close to the Sun.

Thus, the appearance of gravitational waves, on the earth's oceans, suggests that larger Stellar Cores are now approaching the Earth and that they started doing so, in 2017.

But, if these objects are causing sea level changes on the earth's oceans, what effect would they have when they move over land? For now, we are not likely to see the land buckle up and down, as one of these objects passes overhead, although the time for that may still come. But, these objects will create regions of changed gravity, on the surface of the earth, a region will be pulled toward the center of the earth, with a greater force than before, and an adjacent region will be pulled, with a lesser force than before, and as the object moves, the regions affected change, so that the effect is like that of a wave of alternating greater than normal and less than normal density, moving across the ground, i.e. like a p seismic wave. This will cause the ground to break and fissure, due to being alternatively pulled and pushed, in different directions. It will cause hills to break open and slide downwards.

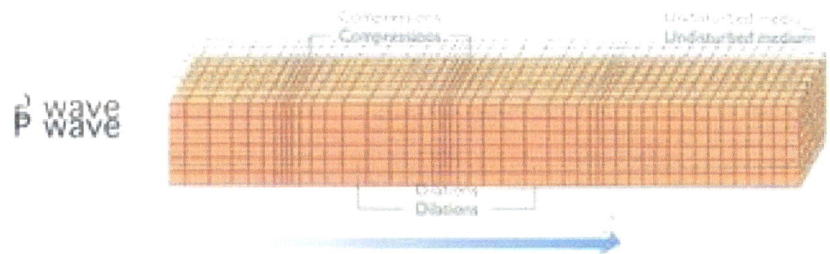

Figure 4:10: A Stellar Core, which is able to create a tidal wave on the ocean, will affect the ground also. The effect will be like what occurs when a p seismic wave passes through the ground, it will create alternating regions of high and low density, thus causing the ground to break up.

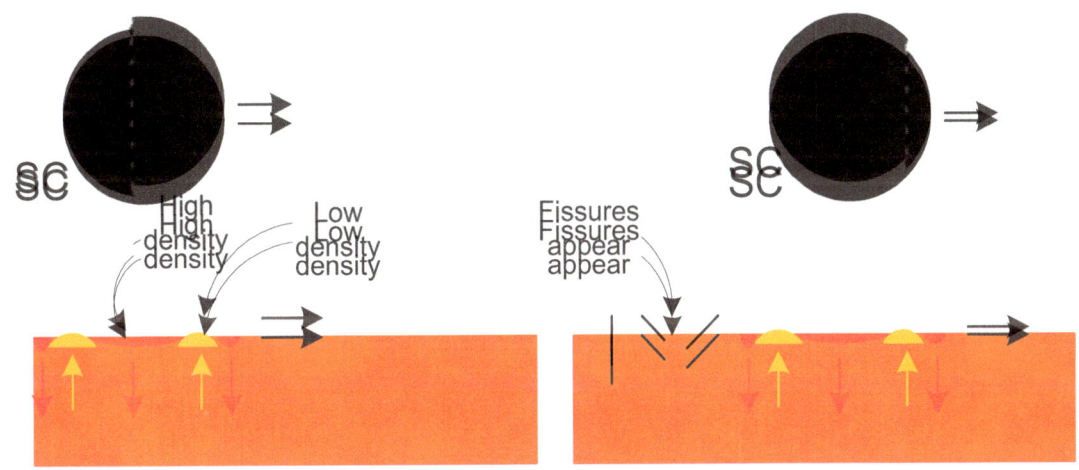

Figure 4:11: A large Stellar Core, outside the atmosphere, but very close to the earth, which is able to generate gravitational waves on the ocean, will create a moving gravitational wave, on the surface of the earth, creating alternating regions of higher and lower density, which will cause the ground to break and fissure.

The reason why larger Stellar Cores are approaching the earth may be due to the Sun having weakened, so that the earth's ability to generate energy has now become an attraction, for some of these objects. This would mean that larger and larger objects are likely to approach earth, over time, thus causing increasingly more severe tidal waves and fissuring of the earth's surface.

In conclusion, larger Stellar Cores seems to now be reaching the earth. These objects produce tidal events, in the earth's oceans, and will, therefore, produce a slowly moving seismic wave, on land, which will break-up the ground and cause it to the fissure.

References:

[1] Albers, C. (2018). Article 343: Hurricane Florence: a cover-up for the tidal force from an object in outer space.

[2] Albers, C. (2018). Article 345: Hurricane Florence tidal event and gravity waves

[3] https://weather.com/storms/hurricane/video/storm-surge-and-flooding-threat-from-hurricane-florence.

[4] Albers, C. (2018). Article 227: Stellar Cores affecting the earth and possible connection to Volcanic Eruptions (in Book 6: Planet X Physicist Articles: Part 1).

[5] Albers, C. (2018). Article 188: What is causing the ocean to recede all over the world? (in Book 3: Planet X Reveals Gravity and Light)

[6] Albers, C. (2018). Article 361: Planet X producing gravitational vortices in the earth's atmosphere since 1643.

[7] Albers, C. (2018). Article 348: Venus bulge: gravitational waves and hollow planets (in Book 8: Planet X and the Solar System).

Chapter 5

370. Hurricane Michael used to cover up tidal event due to Planet X approach

The Youtube channel inTruthbyGrace has done a good job of reporting that Hurricane Michael was not even close to being a hurricane, on October 8th and October 9[th,] 2018. The cloud pattern associated with the storm, as seen on radar, started curving into a vortex shape, on October 8th, indicating that it was artificially formed. However, on October 9th, only half of the vortex had formed. In addition, the winds recorded by ships and buoys, in the areas affected by the hurricane, were too low for the storm to be categorized as a hurricane, as maximum sustained speeds were no greater than 60 miles per hour, at the time it was located near Cuba. A category 1 hurricane has to have a minimum recorded sustained wind speed of 74 mph. The same Youtube channel also showed that on October 9th, when the supposed eye of the storm was north of Cuba, the winds recorded by buoys, in the area, did not show that sustained wind speeds had reached a speed, which would indicate that this storm was anywhere near to a category 1 hurricane, much less a category 2 or 3 as it was being reported.

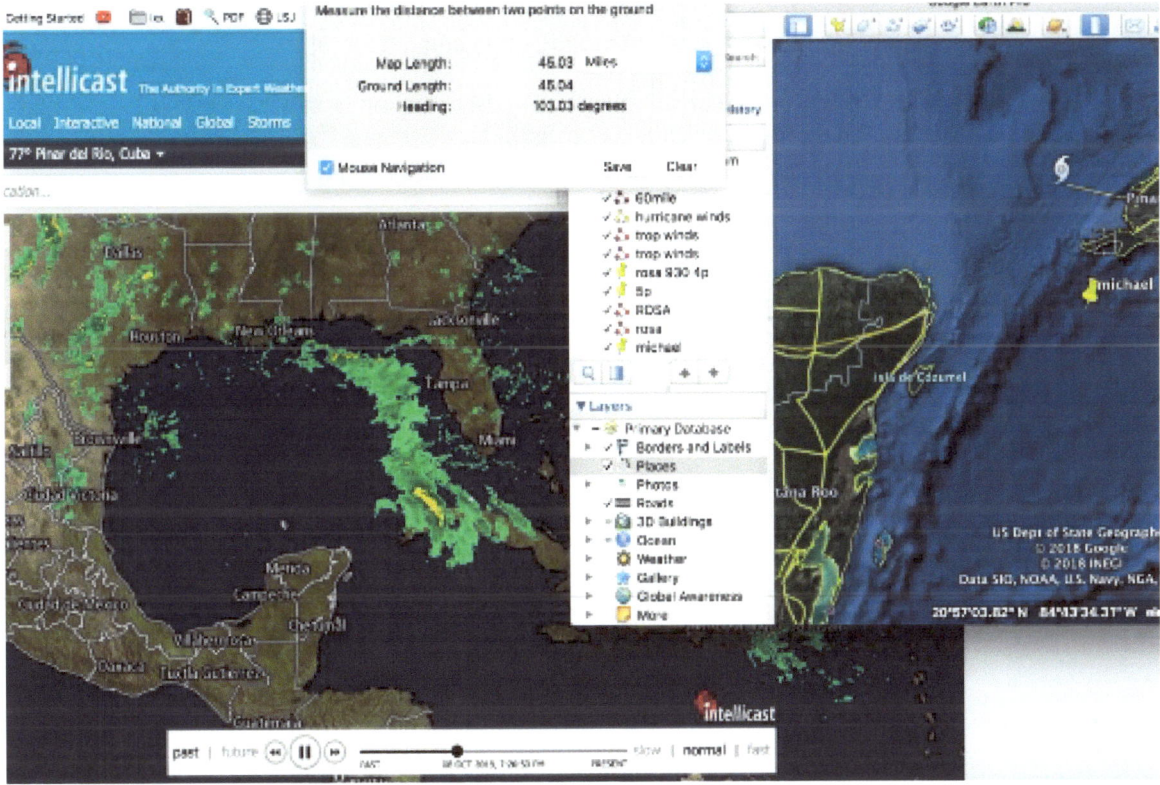

Figure 5.1. Screenshot from Youtube video by inTruthbyGrace from October 8[th,] 2018 showing that on the radar, the storm did not look like a hurricane because no eye and vortex were visible but that some rotation had started [1].

Figure 5.2: Table showing wind speeds associated with different hurricane categories. A category 1 hurricane must have a wind speed of at least 74 mph (64 kts) and a tropical storm must have a wind speed of at least 39 mi/h (35 kts). This means that Hurricane Michael, on October 8th was no more than a tropical depression.

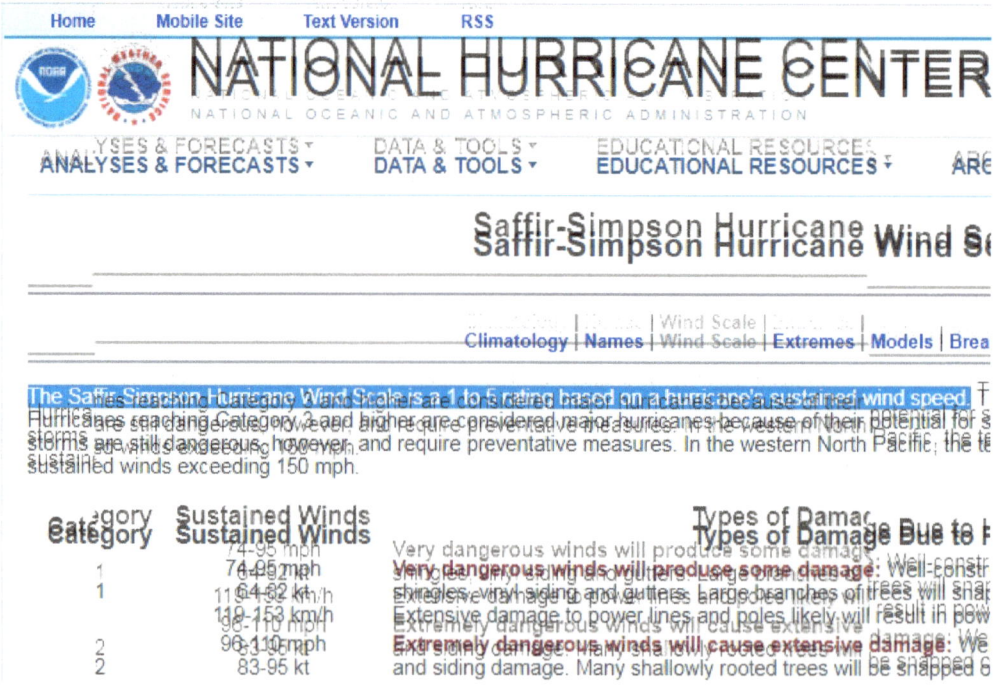

Figure 5.3: Screenshot of the NOAA webpage stating that a hurricanes rating is based on sustained wind speed, i.e. not wind gusts [2].

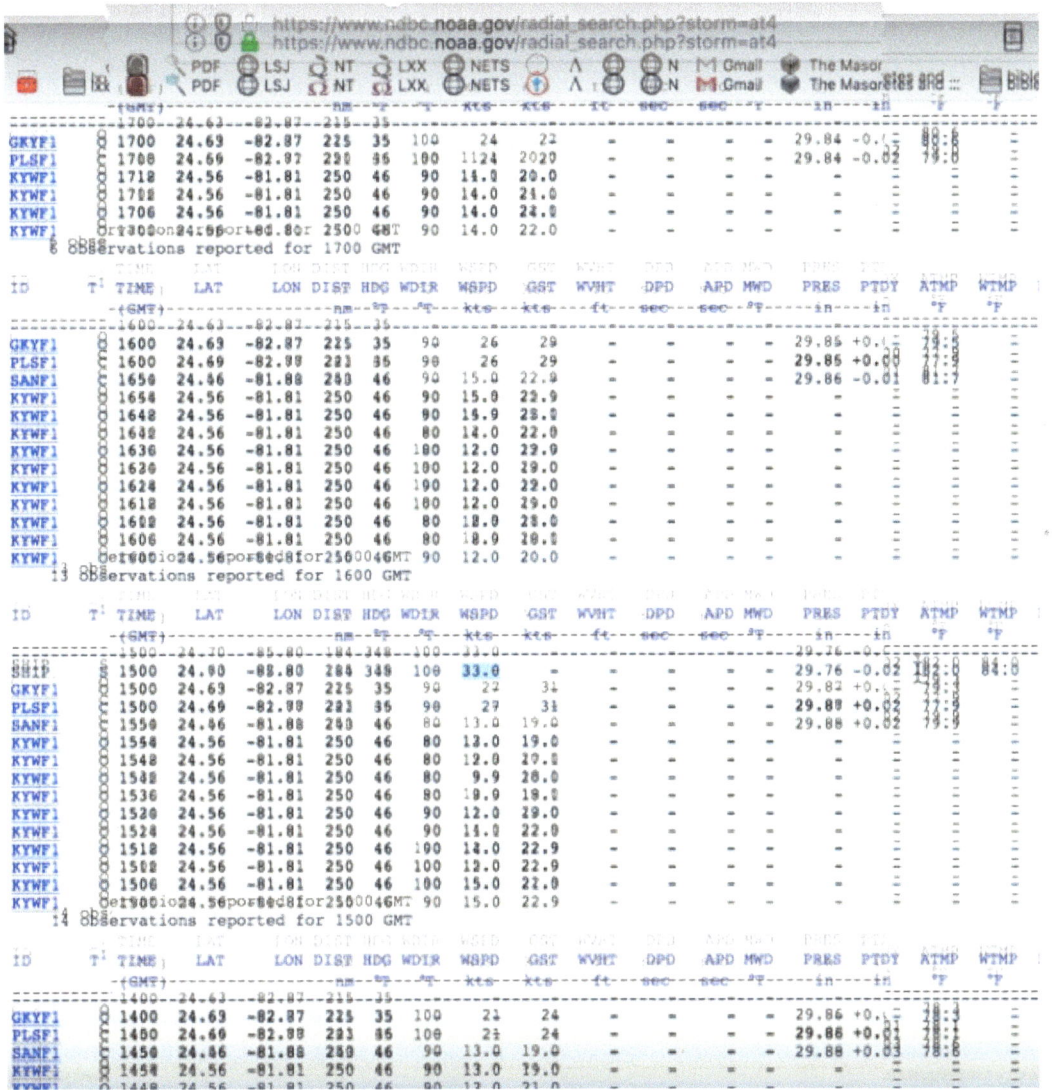

Figure 5.4: Another screenshot from the October 8th, 2018 video, by inTruthbyGrace, showing that the maximum sustained wind speed measured in the area affected by the 'hurricane,' was what was measured by a passing ship, like 33 knots, which translates into 38 mph [1]:

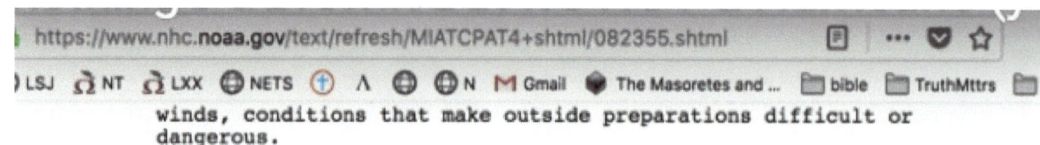

Figure 5.5. NOAA was reporting that Michael was a tropical storm and forecast to become a major hurricane (category 2 to 4 at least) by Tuesday night October 8th. How is that possible? The buoy data showed that it was only a tropical depression. How can the reconnaissance aircraft measure something that is not corroborated by the buoy data or the radar images? High altitude winds, as measured by an aircraft, cannot be the cause of damage on the ground, only surface winds can cause damage on the ground, and only surface winds can produce the waves that lead to a storm surge occurring. The buoy data clearly indicates that wind speed never went close to being what would be expected from a tropical storm, on October 8th, and there was definitely no cloud vortex to indicate that any hurricane, or any cyclonic storm, had formed. So how can they forecast that this was going to become a major hurricane?

The only possible reason, they can do that, is because they were planning to manufacture the hurricane. In the case of Hurricane Florence, the first time that a category 1 hurricane wind speed was measured was when the hurricane reached shore [3]. So it seems that they only have the ability to generate such a storm close to land, possibly because that is where the radar (NEXRAD) stations are, which they seem to use to energize and steer these storms. This means that these are not real hurricanes; these are artificially created storms, which only take shape when they are supposed to be very near to shore. The rest of the time, they are an imaginary fabrication by NASA and NOAA.

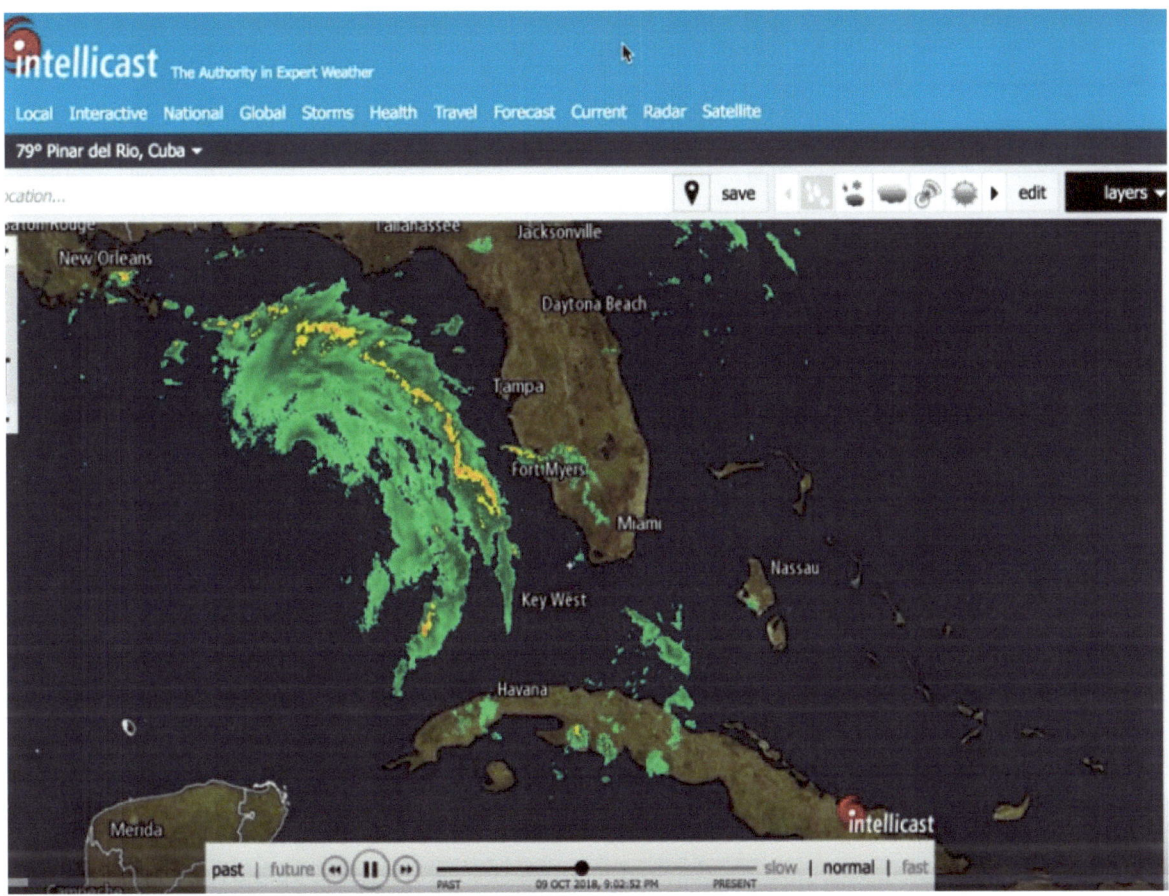

Figure 5.6. Screenshot from video by inTruthbyGrace, from October 9th, 2018, at 9:02 pm, showing a radar image of the storm [4]. Notice that the full vortex had not formed. This seems to be a forming an artificial storm system with the potential for cyclonic circulation. A major hurricane was supposed to form by Tuesday night, October 9th though!

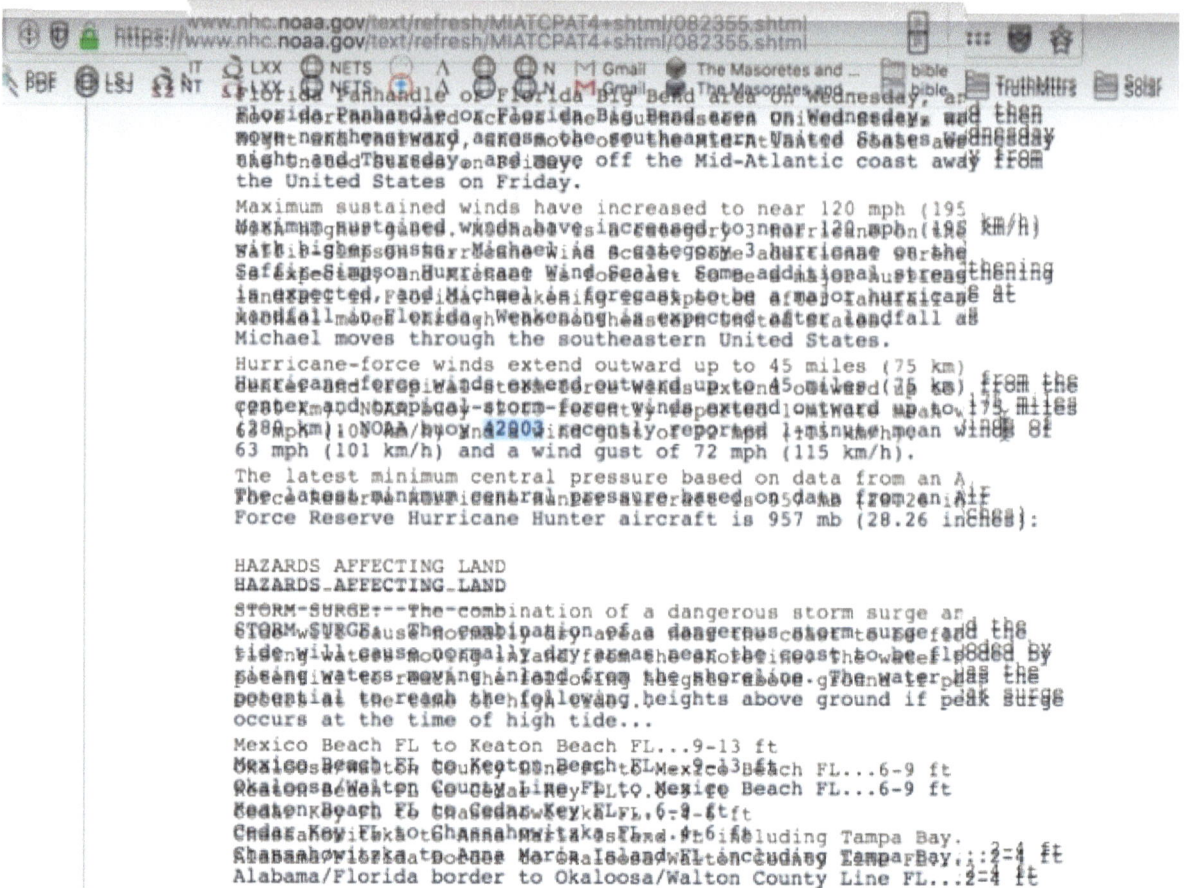

Figure 5.7: NOAA categorically states that Michael is a category 3 hurricane on October 9th, 2018 (from video by inTruthbyGrace) and has sustained wind speeds of 120 mph and is supposed to strengthen further by landfall. They then quote buoy data from buoy 42003, as a 1 minute average, indicating that they use the buoy data to track and categorize hurricanes. But, the buoy data measurements are only 10 minutes apart, so, a 1 minute means, means that they took the highest wind gust, and sustained wind speed measurements, of the day, and averaged them. In other words, they manipulated data to make it appear worse than it was. But even then, even with the manipulation, this buoy data, the 1 minute mean gives only a 63 mph wind speed. The wind gust does not count but even the wind gust speed is only 72 mph. A category 1 hurricane has to have sustained wind speeds of at least 74 mph.

So a major hurricane was supposed to form by this time (Tuesday night October 9th) but no hurricane appears on radar or from the buoy data. We are supposed to have a category 2 or 3 hurricanes but we only have a tropical storm? NOAA categorically states that it is a category 3 but the data shows that they are clearly lying.

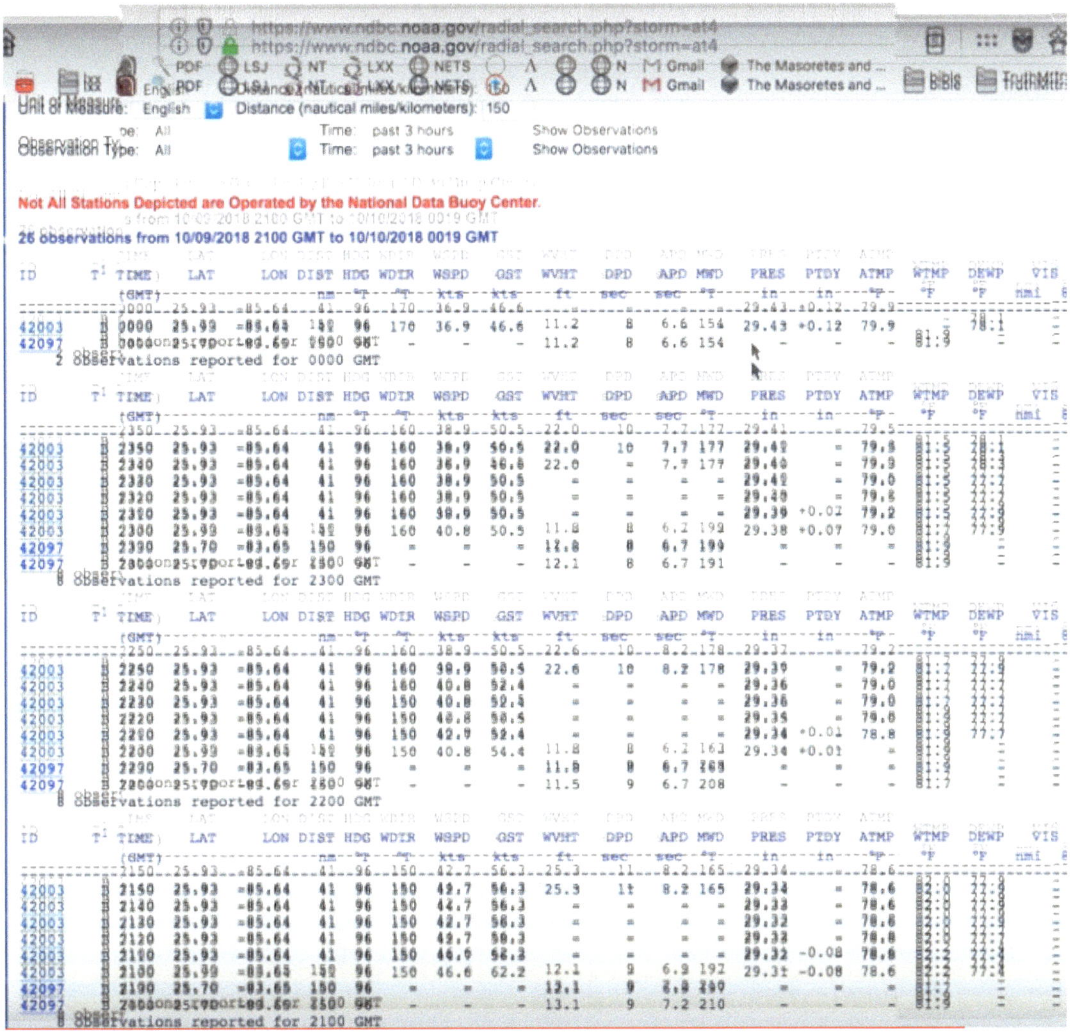

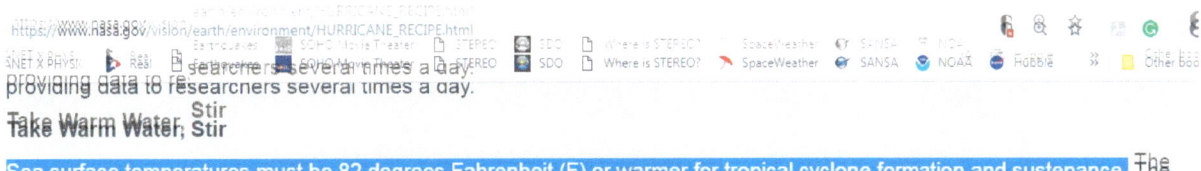

Figure 5.8: Buoy data from 9:00 pm onwards, on October 9th, 2018, indicates that the sustained wind speeds never went over 46.6 kts (53.6 mph), but it has to be at least 64 kts (74 mph), in order for the storm to be categorized as a category 1 hurricane. This means that this was no more than a tropical storm (wind speed of at least 35 kts). NOAA is clearly lying. The higher wind speed was recorded when the water temperature was measured to be 82.2°F, but as the water temperature decreased the wind speeds dropped as is expected as the water temperature has to be at least 82°F in order for a tropical cyclone to form or be sustained.

Figure 5.9: Screenshot of the NASA webpage stating that sea surface temperatures must be 82°F, and above, in order for a tropical cyclone to form, and be sustained [5]. The temperatures as recorded by the buoys dropped as the night progressed to below 82°F, so how could the storm have been sustained overnight?

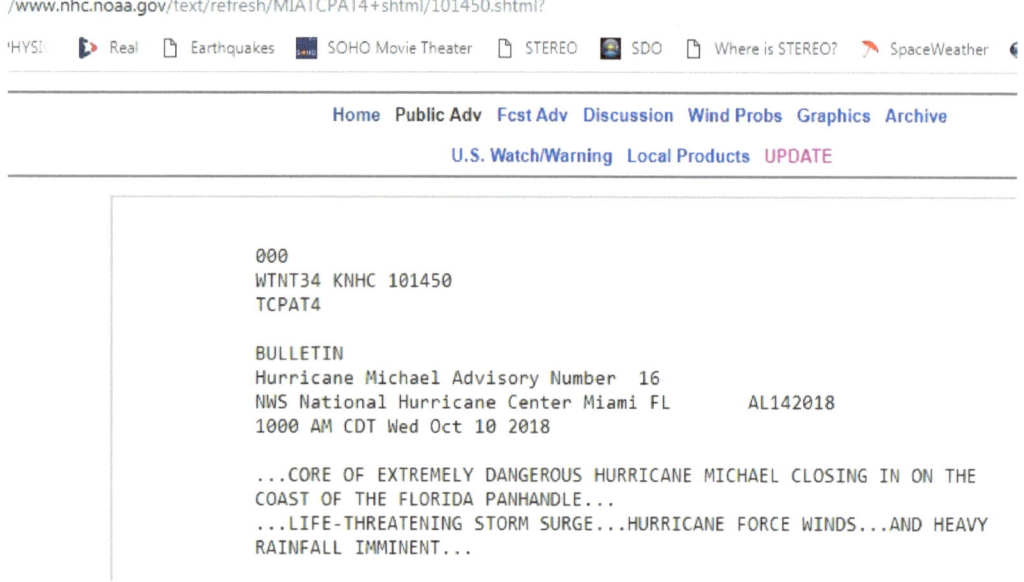

Figure 5.10. NOAA states in the 10 am a bulletin on October 10th that the core of extremely dangerous hurricane Michael is closing in on the Florida Panhandle.

So, how could NOAA report that an extremely dangerous hurricane was approaching the Florida coast, when at midnight, October 10th the storm could not be classified as anything but a tropical storm? And in addition, the water temperature, at least until midnight was too low to sustain the storm? The water temperature would be likely to drop further after midnight and until sunrise which should make the sustainability of this storm even worse.

But why would they manufacture a storm and lie about its strength in such a blatant manner? It seems to be all about the storm surge. They know a tidal surge is going to occur, they cannot stop it but they can try and justify it with a fake artificially produced hurricane.

```
nhc.noaa.gov/text/refresh/MIATCPAT4+shtml/101450.shtml?

    Real    Earthquakes    SOHO Movie Theater    STEREO    SDO    Where is STEREO?    SpaceWeather

            hurricane on the Saffir-Simpson Hurricane Wind Scale. Some
            strengthening is still possible before landfall. After landfall,
            Michael should weaken as it crosses the southeastern United States.
            Michael is forecast to become a post-tropical cyclone on Friday, and
            strengthening is forecast as the system moves over the western
            Atlantic.

            Hurricane-force winds extend outward up to 45 miles (75 km) from the
            center and tropical-storm-force winds extend outward up to 175 miles
            (280 km).  A private weather station at Bald Point, Florida,
            recently reported a sustained wind of 54 mph (87 km/h) with a gust
            to 61 mph (98 km/h).  A wind gust to 46 mph (74 km/h) was recently
            reported inland at Tallahassee, Florida.

            The latest minimum central pressure based on data from the
            reconnaissance aircraft is 928 mb (27.41 inches).


            HAZARDS AFFECTING LAND
            ----------------------
            STORM SURGE:  The combination of a dangerous storm surge and the
            tide will cause normally dry areas near the coast to be flooded by
            rising waters moving inland from the shoreline. The water has the
            potential to reach the following heights above ground if peak surge
            occurs at the time of high tide...

            Tyndall Air Force Base FL to Aucilla River FL...9-14 ft
            Okaloosa/Walton County Line FL to Tyndall Air Force Base FL...6-9 ft
            Aucilla River FL to Cedar Key FL...6-9 ft
            Cedar Key FL to Chassahowitzka FL...4-6 ft
            Chassahowitzka to Anna Maria Island FL including Tampa Bay...2-4 ft
            Sound side of the North Carolina Outer Banks from Ocracoke Inlet to
            Duck...2-4 ft
```

Figure 5.11. Further down in this report from October 10th we are told that sustained wind speeds of 54 mph were recorded consistent with this storm being only a tropical storm, not a category 1 hurricane (minimum is 74 mph) at all. So how can they forecast a maximum storm surge of 9 to 14 feet?

Notice that, with hurricane Florence, the storm surge was initially forecast to reach a maximum of 10 feet, at Wilmington beach, when NOAA was stating that it would impact the coast as a category 3 hurricane [3]. The hurricane turned out to barely be a category 1, and that only, once it reached the shore, and yet the storm surge prediction, was never revised as the category of the hurricane was. Not only that, a storm surge even higher than that was confirmed at New Bern; it reached 7 to 11 feet. How is this possible? They simply needed to manufacture the hurricane in order to justify the tidal surge which they knew was going to occur.

Figure 5.12: Weather channel stating that a storm surge of 7 to 11 feet has been verified at New Bern at on September 14th, 2018, the day that hurricane Florence reached shore [6].

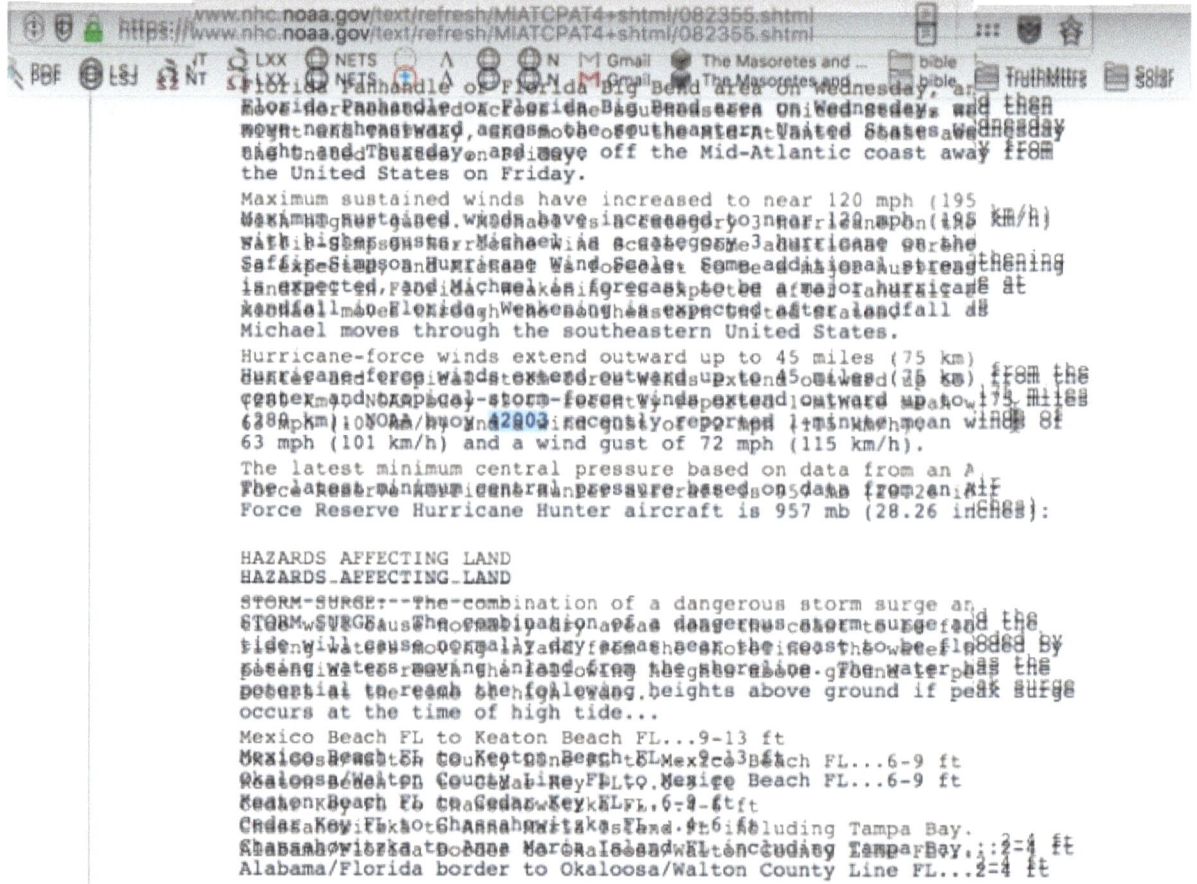

Figure 5.13: October 9th bulletin shows the maximum forecast storm surge to be 9 to 13 feet. This was when they had falsely classified Michael as a category 3 hurricane.

Thus, it seems that these manufactured storms are a cover up for tidal surges. These are not storm surges; these are tidal surges, which they are trying to justify with artificial hurricanes. These hurricanes are so artificial that at times they only exist in the imagination of NOAA and NASA. Tides are produced by outside gravitational influences on the planet. Normal tides are caused by the gravitational attraction of the moon. These abnormal tides, which NASA and NOAA are trying to hide with manufactured storms, are caused by objects, Stellar Cores belonging to the Planet X System, which are approaching the earth. This started in 2017, with the unprecedented water recession events and tidal events which started occurring around the world.

Figure 5.14: Left: Ocean recedes leaving boats sitting on mud, in the harbor in Punta del Este, Uruguay, on August 11th, 2017. The ocean came back but this extreme low tide had never happened before. Right: An empty beach, due to the ocean receding, from the Brazilian coast, on August 12th, 2017, no large storms or hurricane could be blamed for the phenomenon, as there were no storms or hurricanes anywhere near this coastline. This too was unprecedented [7].

Figure 5.15: The ocean floods land in Durban South Africa in March of 2017 indicating that a tidal surge was occurring similar to what artificial hurricane Florence and artificial hurricane Michael are being used to cover up. It was reported at the time that a hurricane off the coast of Madagascar (way too far to affect the Durban coast) was the cause but that is a ridiculous attempt at covering up a tidal event that can only be caused by an object coming from space [7].

They are using artificial cloud seeding to generate a lot of rain and causing the air mass saturated with rain, and with metallic cloud nucleation centers inside it, from chemtrails, to rotate by hitting them with radar, i.e. intense radio waves, in a certain pattern. The excessive rain, over land, is used to cover up the fact that the gravitational wave associated with the Stellar Core approaching the surface of the earth, at this region, would also affect lake and river levels, in addition to the sea level. This level is set by the gravity over this region and these objects exert strong differential gravitational (tidal) forces because they are most likely larger and stronger than the ones that had been approaching earth before. This means that all this damage to property and life which they are artificially creating is only because they are desperate to hide what is going on in the Solar System from the earth's population.

In conclusion, 'hurricane' Michael is an artificial hurricane produced to cover up the approach, to earth, of Planet X Objects, more gravitationally powerful than the ones that had been approaching earth, before 2017. These objects are closely approaching the surface of the earth and exerting strong gravitational forces, on the earth, as a result.

References:

[1] Youtube video by inTruthbyGrace: https://www.youtube.com/watch?v=NVT6uNgnpyw
[2] NOAA webpage: https://www.nhc.noaa.gov/aboutsshws.php
[3] Albers, C. (2018). Article 345: Hurricane Florence tidal event and gravity waves.
[4] Youtube video by inTruthbyGrace: https://www.youtube.com/watch?v=0FPqocy_KnU&t=751s
[5] NASA webpage: https://www.nasa.gov/vision/earth/environment/HURRICANE_RECIPE.html
[6] Albers, C. (2018). Article 343: Hurricane Florence: a cover-up for the tidal force from an object in outer space.
[7] Albers, C. (2018). Article 364: Planet X causing the earth's surface to break up.

Chapter 6

371. Hurricane Michael: tidal event due to an object approaching from space

In Article 370: Hurricane Michael used to cover up tidal event due to Planet X approach [1], I detailed how when hurricane Michael was being reported as a tropical storm, close to Cuba, on October 8th, 2018, it was no more than a tropical depression, or simply, a rain storm, which seemed to have been produced in the area, through most likely cloud seeding (chemtrails). On October 9th, the storm was reported to be a category 3 hurricane, but the sustained wind speeds, from buoy data, showed that it was only a tropical storm, as it had not reached category 1 wind speeds. Now, I am going to show that the buoy data continued to show that the wind speeds, associated with this hurricane, close to the region where it made landfall, never reached category 1 strength and that therefore the hurricane did not approach from the ocean, but from above. There were no winds to cause a storm surge so the inundation could only have been caused by a tidal event, or gravitational force, exerted by an object, approaching from space.

Figure 6.1. CNN reporting that the hurricane made landfall at Panama City.

Figure 6.2: CNN reporting from Panama City at 1:00 pm local time (18:00 UTC) that the hurricane had just made landfall.

Figure 6.3: CNN reporting at 1:04 pm CT (18:04 UTC) that the storm surge had already reached 8 feet, at Apalachicola.

Figure 6.6: Apalachicola region experiencing inundation reported being due to storm surge.

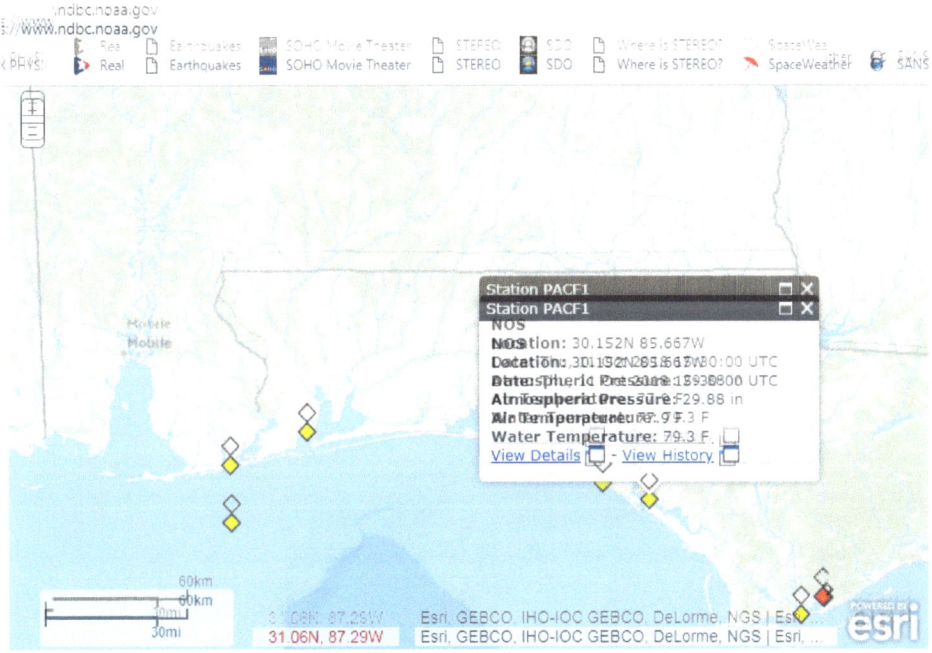

Figure 6.7: The buoy data we will look at: 2 at Panama City (PACF1 and PCBF1) and 1 at Apalachicola (APCF1). The two Panama City (PC) buoys are so close, and the hurricane was reported to be so large, in terms of diameter, that we should expect a large difference in what the two buoys measure.

```
#YY  MM DD hh mm WDIR WSPD GST WVHT DPD APD MWD  PRES  ATMP WTMP DEWP VIS PTDY TIDE
#yr  mo dy hr mn degT m/s  m/s  m   sec sec degT  hPa  degC degC degC nmi hPa  ft
2018 10 11 15 18  MM   MM   MM   MM   MM  MM  MM 1011.8 25.4 26.3  MM   MM  MM  MM
2018 10 11 15 12  MM   MM   MM   MM   MM  MM  MM 1011.9 25.3 26.3  MM   MM  MM  MM
2018 10 11 15 06  MM   MM   MM   MM   MM  MM  MM 1011.8 25.2 26.3  MM   MM  MM  MM
2018 10 11 15 00  MM   MM   MM   MM   MM  MM  MM 1011.8 25.3 26.3  MM   MM +1.3 MM
2018 10 11 14 54  MM   MM   MM   MM   MM  MM  MM 1011.8 25.1 26.3  MM   MM  MM  MM
```

Figure 6.8. Panama City 1 (PACF1) is not currently showing wind speed data but we can see which columns show what: First 3 show date, 4th and 5th show time, 6th shows wind direction, 7th shows wind speed, 8th shows wind gust speed, and the 13th shows air pressure. Time is UTC or universal time. This is the data we will need. The data is in m/s for speed. Standard atmospheric pressure, at sea level, is 1000 hPa, anything below represents a low pressure and thus a storm.

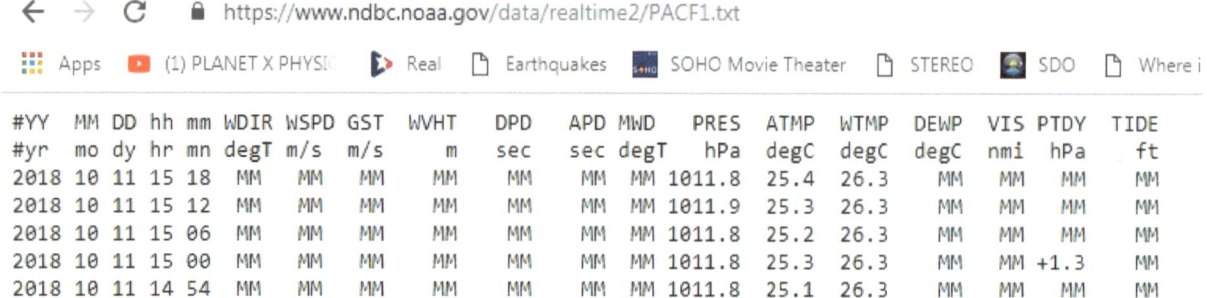

Figure 6.9. Table showing wind speeds associated with different hurricane categories. A category 1 hurricane must have a wind speed of at least 33 m/s. A tropical storm must have a wind speed of 18 m/s. The wind used to classify the hurricane must be sustained wind [1].

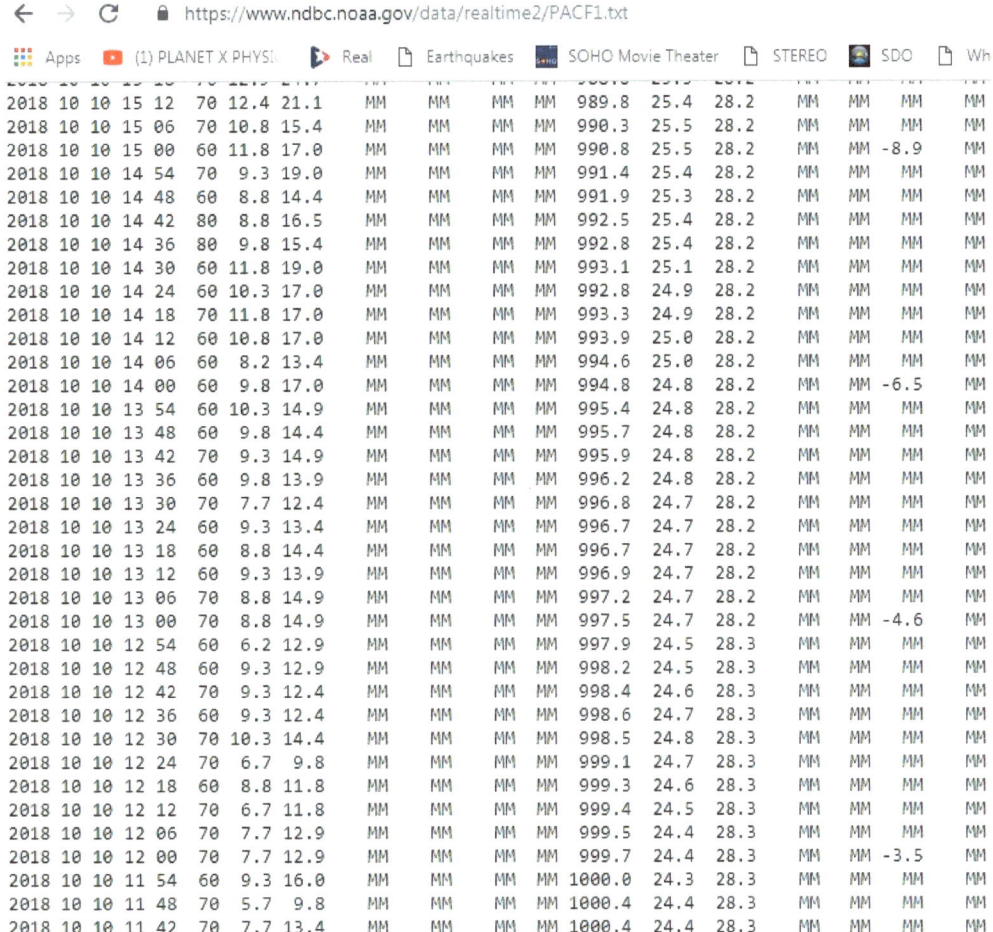

Figure 6.10. Panama City buoy1 (PACF1) data: Pressure started dropping at 11:54 but the wind speed was only 7.7 m/s and it only rose to 12.4 m/s by 15:12. Pressure continued to drop; lowest pressure shown is 989.8 hPa at 15:12. An approaching hurricane should create a drop in air pressure but there should be accompanying increases in wind speed.

Figure 6.11. Panama City Buoy1 (PACF1) data: air pressure minimum: 937.5 hPa reached at 17:48, just before the hurricane made landfall at 18:00 UTC.

2018	10	10	17	36	MM	MM	MM	MM	MM	MM	MM	941.0	24.0	28.3	MM	MM	MM	MM
2018	10	10	17	30	MM	MM	MM	MM	MM	MM	MM	942.6	24.0	28.3	MM	MM	MM	MM
2018	10	10	17	24	MM	MM	MM	MM	MM	MM	MM	944.4	23.9	28.3	MM	MM	MM	MM
2018	10	10	17	18	20	31.9	47.9	MM	MM	MM	MM	947.7	24.1	28.3	MM	MM	MM	MM
2018	10	10	17	12	30	30.4	42.2	MM	MM	MM	MM	952.4	24.1	28.3	MM	MM	MM	MM

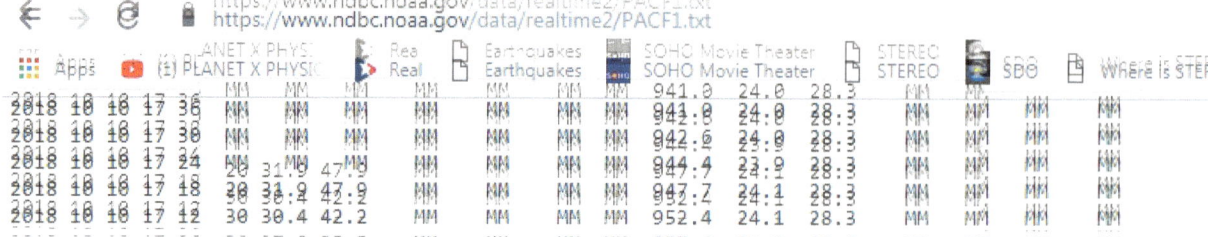

Figure 6.12: The buoy stopped recording wind speed at 17:18, the maximum speed recorded is 31.9 m/s which is too low for a category 1 hurricane. But surely with the width that this hurricane was reported to have had, and it is a category 4, as it reached land, the wind speeds reached, by this time, should have been much higher.

2018	10	10	23	42	MM	MM	MM	MM	MM	MM	MM	1000.3	26.8	26.7	MM	MM	MM	MM
2018	10	10	23	36	MM	MM	MM	MM	MM	MM	MM	1000.3	26.8	26.7	MM	MM	MM	MM
2018	10	10	23	30	MM	MM	MM	MM	MM	MM	MM	1000.0	26.8	26.7	MM	MM	MM	MM
2018	10	10	23	24	MM	MM	MM	MM	MM	MM	MM	999.8	26.8	26.7	MM	MM	MM	MM
2018	10	10	23	18	MM	MM	MM	MM	MM	MM	MM	999.8	26.8	26.7	MM	MM	MM	MM
2018	10	10	23	12	MM	MM	MM	MM	MM	MM	MM	999.6	26.8	26.7	MM	MM	MM	MM
2018	10	10	23	06	MM	MM	MM	MM	MM	MM	MM	999.2	26.8	26.7	MM	MM	MM	MM
2018	10	10	23	00	MM	MM	MM	MM	MM	MM	MM	998.9	26.8	26.7	MM	MM	+11.6	MM
2018	10	10	22	54	MM	MM	MM	MM	MM	MM	MM	998.7	26.7	26.6	MM	MM	MM	MM
2018	10	10	22	48	MM	MM	MM	MM	MM	MM	MM	998.4	26.6	26.7	MM	MM	MM	MM
2018	10	10	22	42	MM	MM	MM	MM	MM	MM	MM	998.0	26.6	26.7	MM	MM	MM	MM
2018	10	10	22	36	MM	MM	MM	MM	MM	MM	MM	997.5	26.4	26.7	MM	MM	MM	MM
2018	10	10	22	30	MM	MM	MM	MM	MM	MM	MM	997.1	26.5	26.7	MM	MM	MM	MM
2018	10	10	22	24	MM	MM	MM	MM	MM	MM	MM	996.8	26.6	26.7	MM	MM	MM	MM
2018	10	10	22	12	MM	MM	MM	MM	MM	MM	MM	996.5	26.6	26.7	MM	MM	MM	MM
2018	10	10	22	06	MM	MM	MM	MM	MM	MM	MM	996.4	26.6	26.7	MM	MM	MM	MM
2018	10	10	22	00	MM	MM	MM	MM	MM	MM	MM	996.2	26.6	26.7	MM	MM	+20.1	MM
2018	10	10	21	54	MM	MM	MM	MM	MM	MM	MM	996.0	26.6	26.8	MM	MM	MM	MM

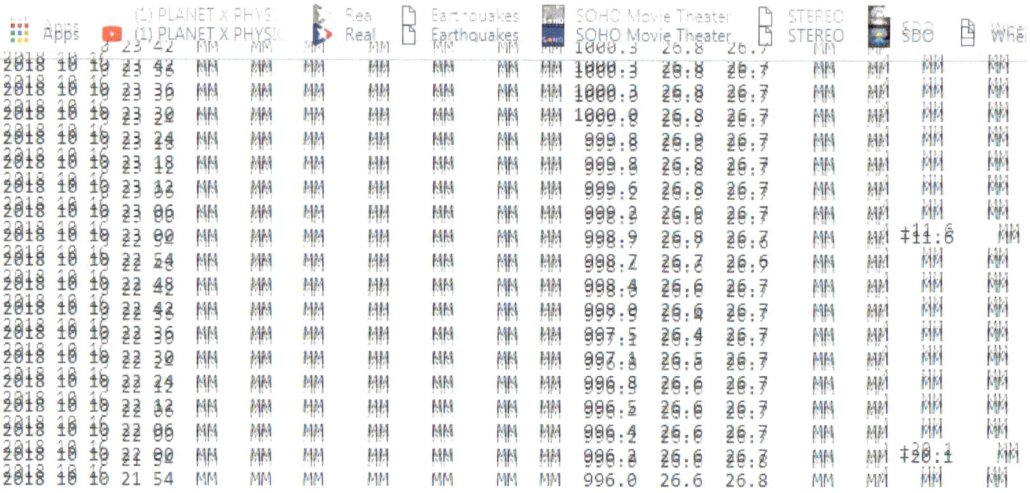

Figure 6.13: Panama City Buoy1 (PACF1) data: air pressure returns to standard at 23:30.

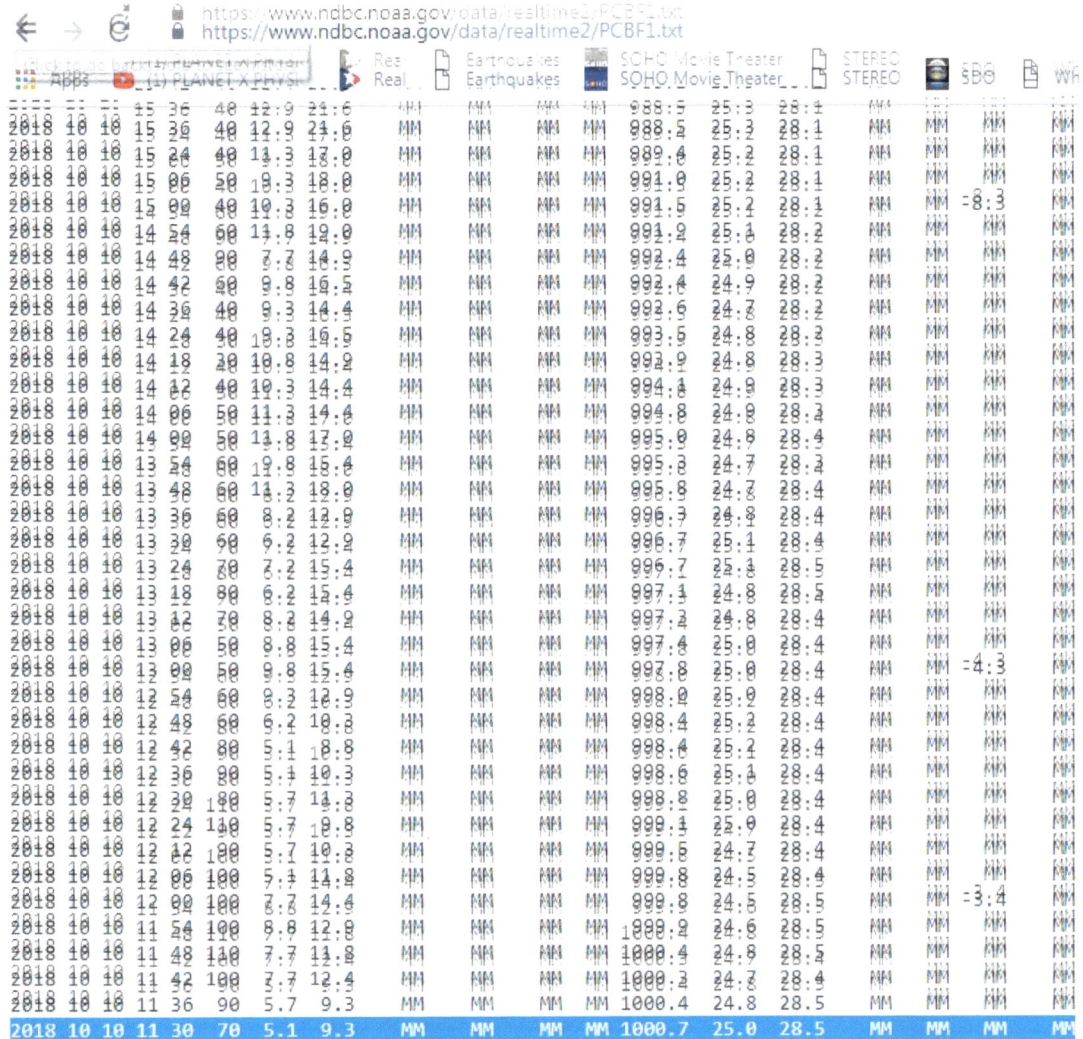

Figure 6.14: Panama City buoy 2 (PCBF1) data: we see something similar to what Panama City Buoy1 data shows, the air pressure starts dropping at 11:54 and drops to 988.5 hPa, by 15:36 (for data shown). The maximum speed reached was 12.9 m/s, way below what we would expect with such a large hurricane over the area.

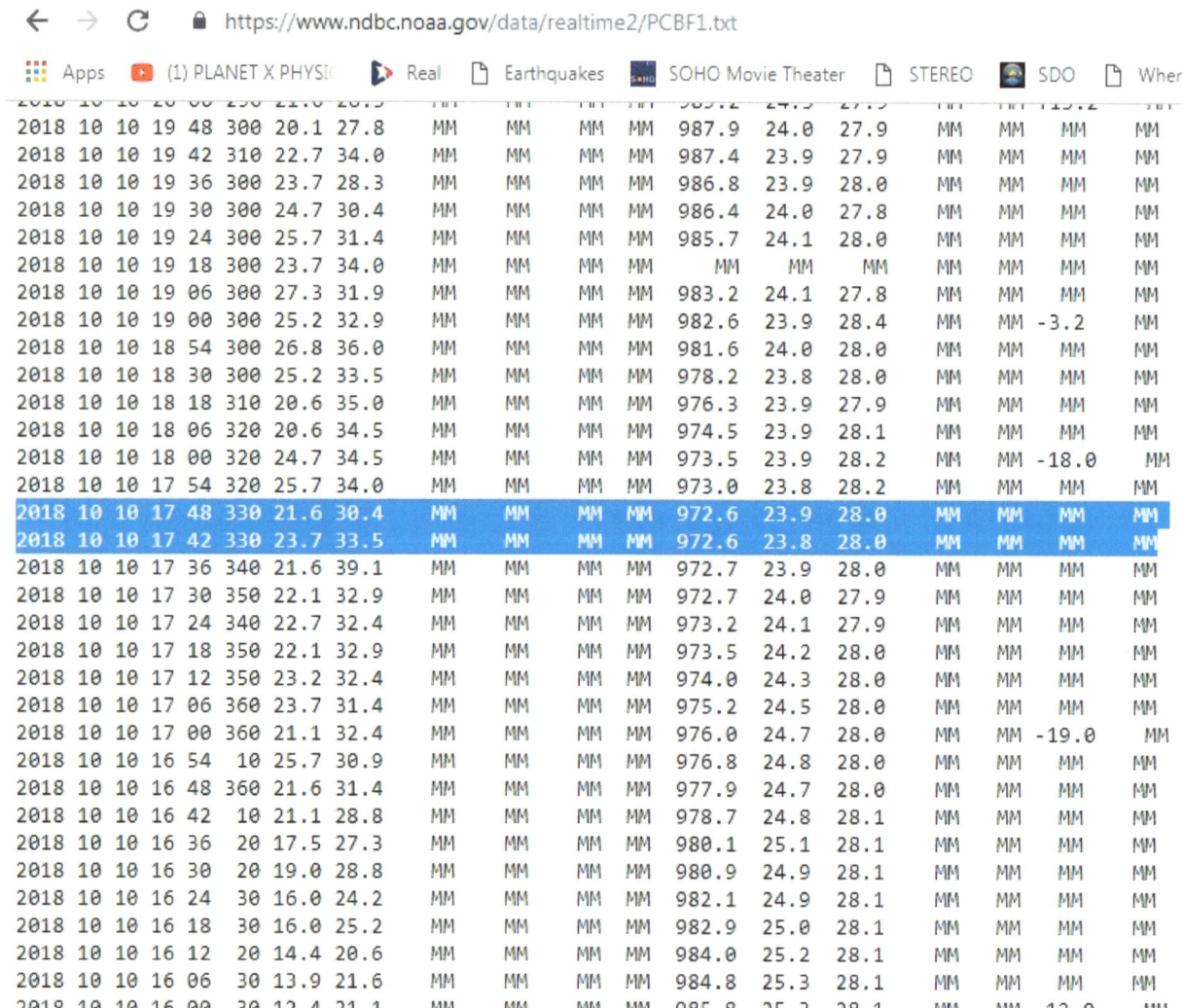

Figure 6.15. Panama City (PC) buoy 2 (PCBF1) data: minimum air pressure reached: 972.6 hPa, at 17:42 (6 minutes before PC bouy1). This is not as low as PC bouy1 minimum air pressure reached, which was 937.5 hPa, suggesting PC buoy1 was closer to the center of the storm. Maximum wind speed: 27.3 m/s and after minimum pressure was reached, at 17:54. Minimum air pressure was reached only 6 minutes before the hurricane was reported to have made landfall, and thus at about the same time. Maximum wind speed reached was below category 1 strength, inconsistent with a category 4 hurricane passing over the area. The hurricane was so wide it should have started impacting the area hours before landfall.

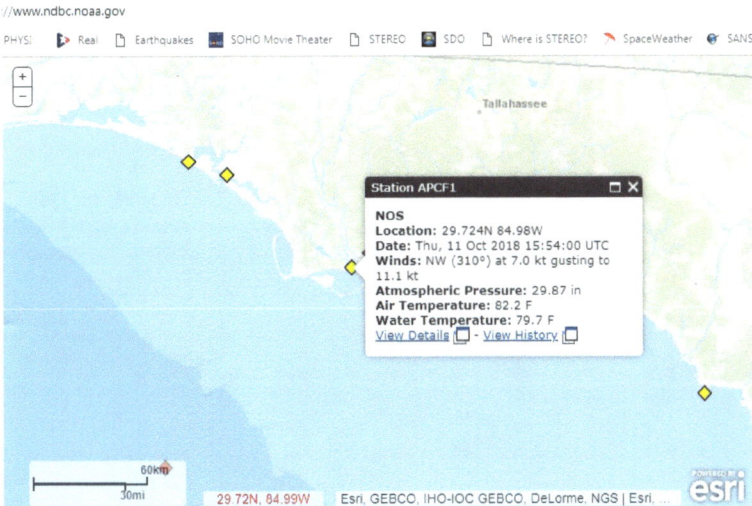

Figure 6.16. Apalachicola buoy: Apalachicola was where the strongest inundation seems to have occurred.

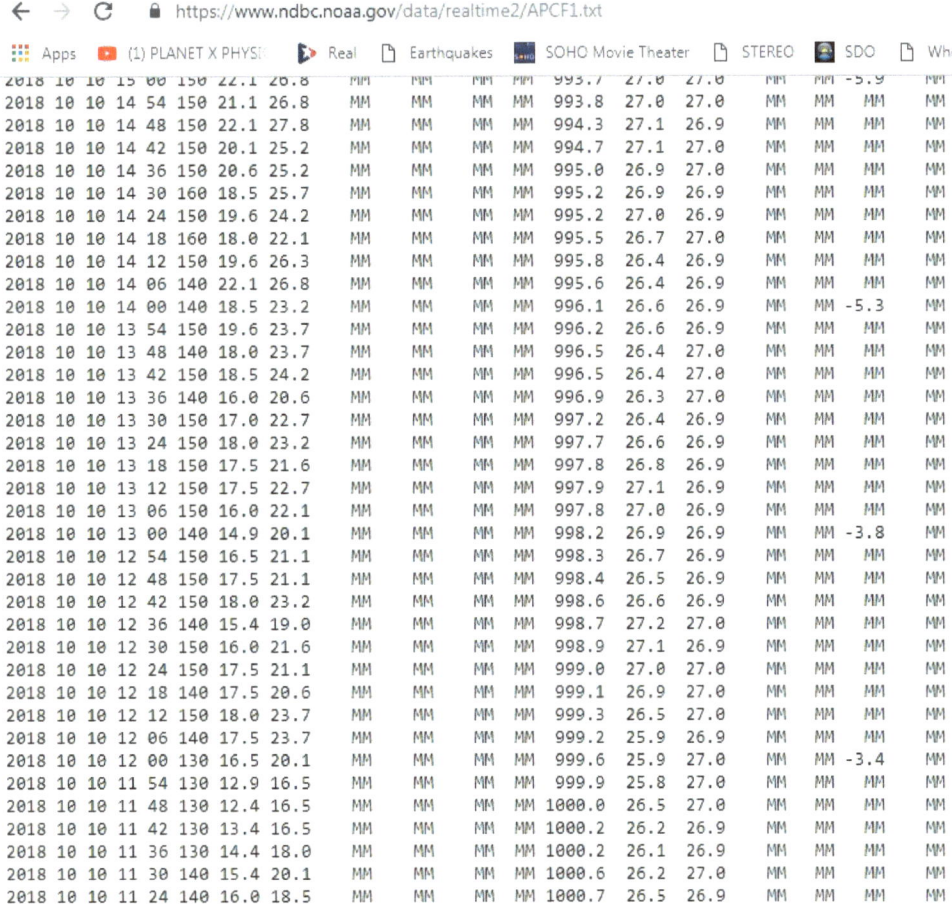

Figure 6.17. Apalachicola buoy (APCF1) data: Wind speed never reached hurricane force (33 m/s), maximum reached was 22.1 m/s at 15:00. Air pressure began dropping at 11:54, as with other two buoys.

45

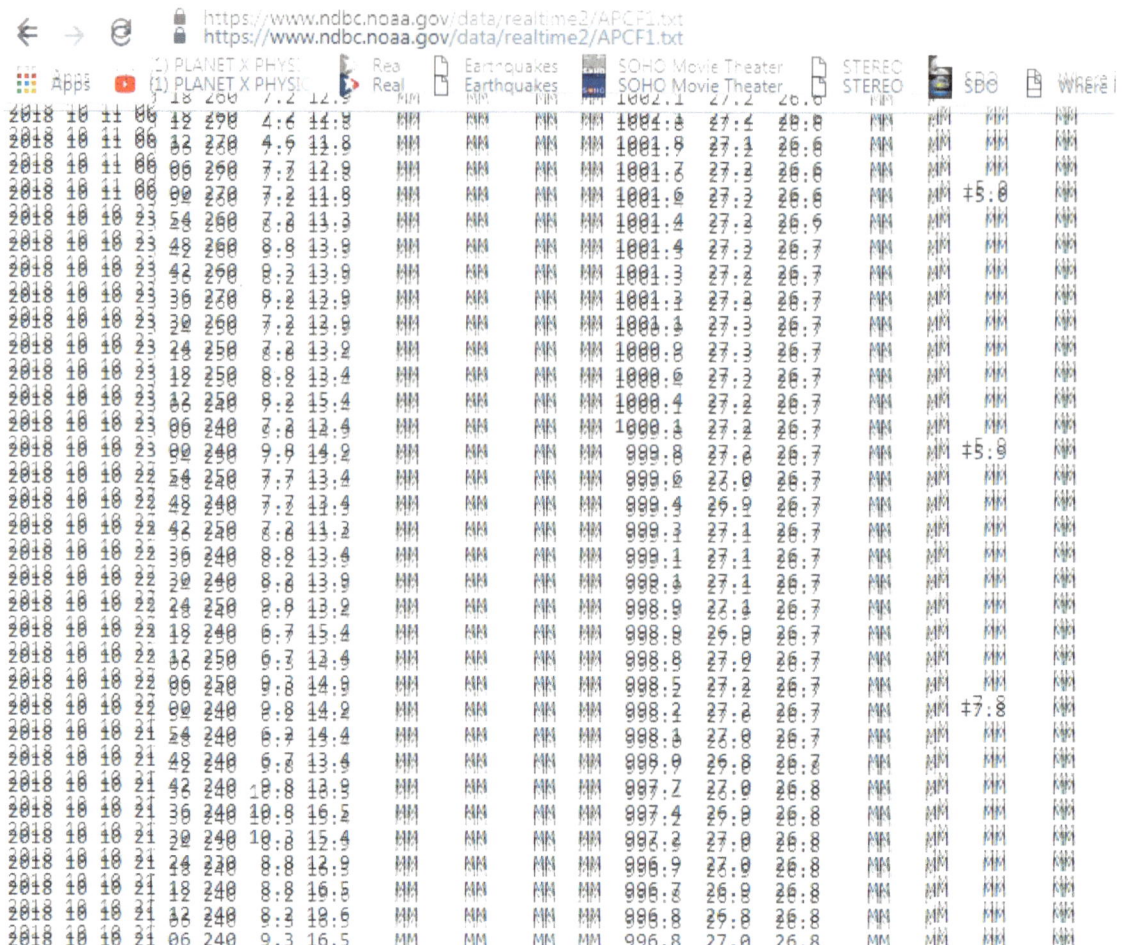

Figure 6.18: Apalachicola buoy (APCF1) data: Air pressure returned to normal at 23:06 as with the other buoys.

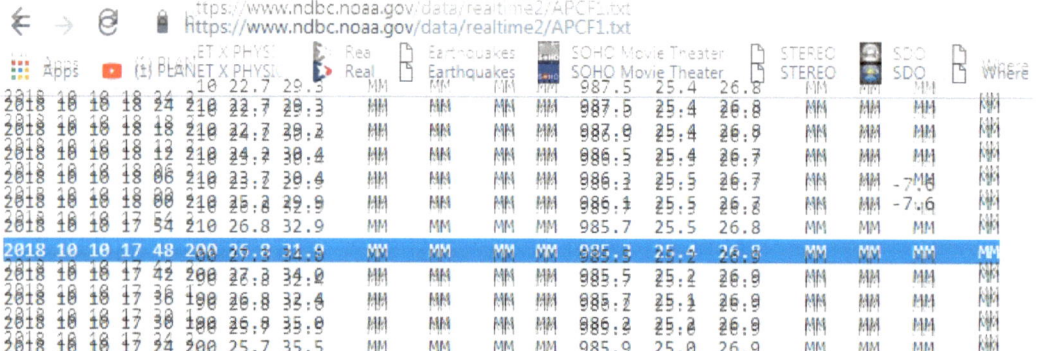

Figure 6.19: Apalachicola buoy (APCF1) data: minimum pressure: 985.3 hPa at 17:48, maximum wind speed reached just before and just after minimum air pressure reached. This is higher than for the other 2 buoys suggesting this buoy was further from the center of the storm. The maximum wind speed reached was 27.3 m/s, the same as PC buoy2.

So how can we make sense of this data? These buoys never showed that a hurricane had passed overhead, as the recorded wind speeds were always below hurricane force.

In addition, the air pressure started dropping at around 11:54 and went back to standard air pressure at around 23:06, for PC buoy1 and Apalachicola, and a little later, for PC buoy2. So, the pressure dropped for about 12 hours, with the landfall time of 18:00 is around the middle of that time interval. But wind speeds never reached hurricane force and were lower in Apalachicola was the inundation seemed to be greatest. There is no evidence that a hurricane passed through the region of the buoys. How can a hurricane make landfall just north of the buoys without it passing the buoys? How can the hurricane have such a huge diameter if it made landfall in Panama City without high wind speeds being recorded by the buoys? This suggests that the hurricane did not come from the ocean it came from above, it landed in the area and it must have had a much smaller diameter than was being reported. The tidal surge had a much higher diameter than the associated storm.

The minimum pressure was recorded at PCbuoy1 so this must have been closer to the center of the storm and the others being closer to the edge.

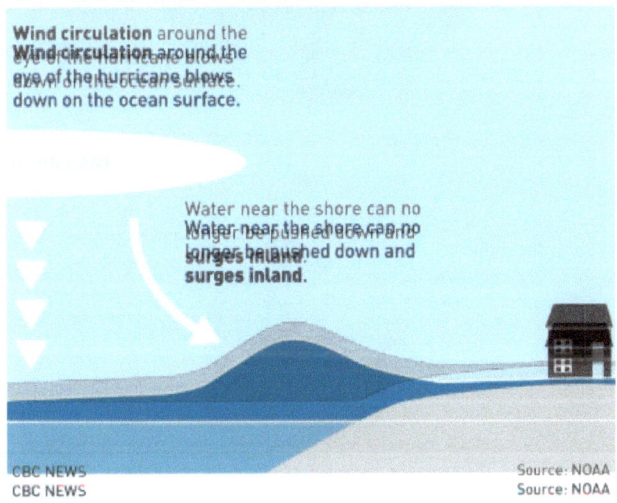

Figure 6.20: NOAA shows that an approaching hurricane wind circulation pushes water downwards, but that close to land, it cannot be pushed down any further and so surges inland.

Well, if that is the case, we would need some evidence that there was a hurricane out to sea or over the region pushing water, causing wind circulation which would then push water onto land, but there is none. The wind speeds never reached hurricane force for any of the regions looked at. Air pressure most certainly cannot push water downwards. If anything, low air pressure should allow water to rise up, because it is pushed down with less force. This means that we had a storm surge, with no hurricane force winds off shore to push it onto land. In other words, there was nothing to cause a storm surge. So this had to be a tidal event.

The data shows very low pressures and the highest wind speeds are reached around the time that the pressure is lowest, even though they are too low to be hurricane force. And much lower than what was reached inland. The reported wind speeds on land were 155 mph (96 m/s), but these winds were never recorded off shore, which means that a hurricane touched down over land. It touched down as a tornado does, from above. The edges of the gravitational field then affected the ocean and that was what was measured by the buoys. The water that surged inland was due to a tidal surge, created by a gravitational anomaly, in the area centered on a point on land.

A normal tide, on earth, is created by the moon. The moon is an external object, as it is outside of the earth's atmosphere. So a tidal surge of this magnitude would have to be created by an object approaching earth, from outside the earth's atmosphere. Its gravitational field would create a low pressure and a corresponding tidal bulge, somewhere under the object. The gravitational force should also give rise to circulating air, in the form of an air vortex.

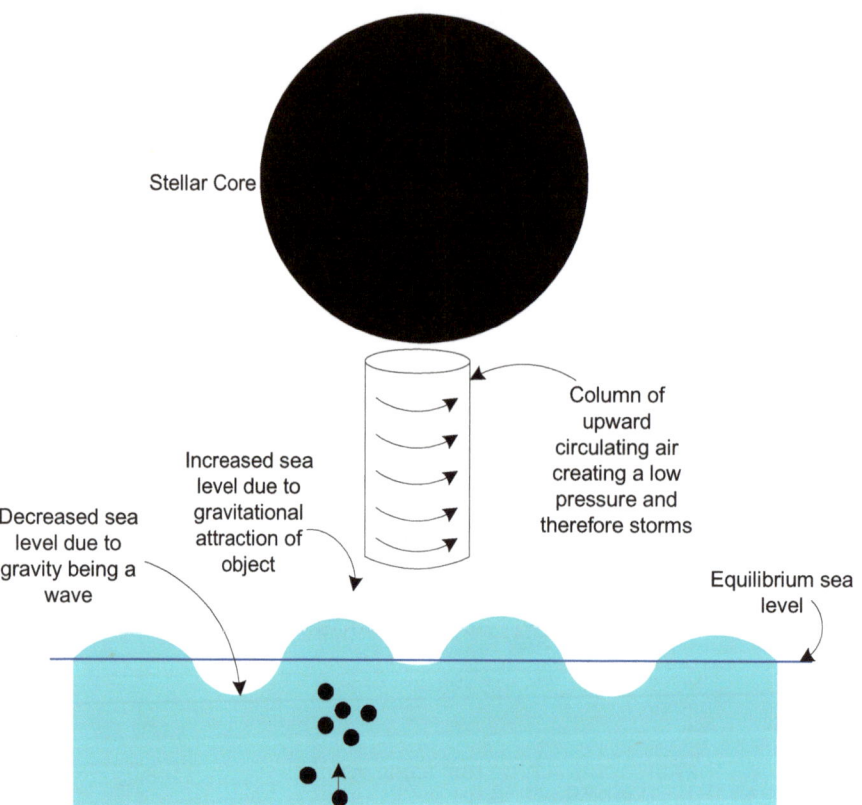

Figure 6.21. The gravitational wave created by an object approaching from outer space. This is consistent with Apalachicola being further from the center of the storm created by an external gravitational field and yet experiencing a greater tidal surge.

Gravity seems to be a stationary wave, with highs and lows, and a central minimum (see Article 349: Planet X: gravitational and electric effects on the Planets) [2]. This is why all gravitational vortices have central hollows. Regions of decreased sea level lead to ocean recession events.

Figure 6.22. A hole at the bottom of a container, filled with water, will cause the earth's gravitational attraction exerted through that hole, on the water, to create a vortex of water.

Figure 6.23. Left: Ocean recedes leaving boats sitting on mud, in the harbor in Punta del Este, Uruguay, on August 11th, 2017. The ocean came back but this extreme low tide had never happened before. **Right**: An empty beach, due to the ocean receding, from the Brazilian coast, on August 12th, 2017, no large storms or hurricane could be blamed for the phenomenon, as there were no storms or hurricanes anywhere near this coastline. This too was unprecedented (see Article 227: Stellar Cores affecting the earth and possible connection to Volcanic Eruptions) [4]. These could only be caused by a tidal force which can only occur when an object closely approaches earth from space.

In conclusion, the hurricane, which affected the region around Panama City, on October 10th, 2018, and which led to a huge amount of inundation, from the ocean, could not have been caused by a hurricane, which formed out to sea, and blew water inland, as the buoy data clearly shows. The inundation, which occurred, had to be caused by a local gravitational anomaly, which only an object closely approaching earth from space can cause.

References:

[1]   Albers, C. (2018). Article 370: Hurricane Michael used to cover up tidal event due to Planet X approach.
[2]   Albers, C. (2018). Article 349: Planet X: gravitational and electric effects on the Planets.

Chapter 7

377. Pink clouds over Florida indicate something is very wrong with the Earth and the Sun

A photograph below of the skies over Florida in the aftermath of Hurricane Michael shows pink skies and a blue cloud. What could possibly cause this type of alien looking sky? In Article 376: What caused the bright pink sky in Florida during the Hurricane Michael event? [1] I showed that the pink skies had to be caused by an object emitting that color of light. Now, after looking at the photograph below, I realized that we did not just have a large object immersed in a cloud and illuminating all the clouds below it. No, the clouds in the sky were the actual sources of the light, some were emitting pink light and some were emitting blue light.

Figure 7.1: A blue cloud in the foreground with pink clouds behind it. This can only arise if the clouds themselves are emitting these colors of light.

These clouds must have formed when a huge amount of Stellar Core dust entered the atmosphere, water condensed around the dust and the energy absorption process, caused the clouds to emit light in different colors, such as the pink and blue, we see here.

The objects, responsible for the strange colors seen in our skies, are members of the Planet X System of Stellar Cores or dead stars, planets, and moons. They are dead in the sense that they are energy depleted. Their core is no longer able to generate energy and the particles making up the object have become depleted in gravitational photon energy. Since it is this energy which allows an object to have gravitational influence and emit light from its atmosphere, they have low gravitational influence and thus shed their outer layers of material, until the core of the object becomes exposed, hence the name Stellar Core. They thus generate a lot of debris and since so many of these objects have invaded the

Solar System and have come to the Sun to absorb energy from it, the inner Solar System is now filled with their debris in the form of dust and other particles in different sizes (see Article 275: Planet X debris field impacting earth) [2]. Thus, huge amounts of dust are continuously arriving at earth, entering our atmosphere, absorbing energy from the earth and creating luminescent clouds in our atmosphere. In addition, actual Stellar Cores, are arriving on earth and creating various effects in our atmosphere such as tornadoes, water spouts and hurricane, which are really gravitational vortices. These too will be immersed in clouds, which will also emit these colors. But the extent and brightness of the bright colors, in the sky, after Hurricane Michael impacted the area, suggests that an abnormally large amount of this Stellar Core dust entered the atmosphere during this event.

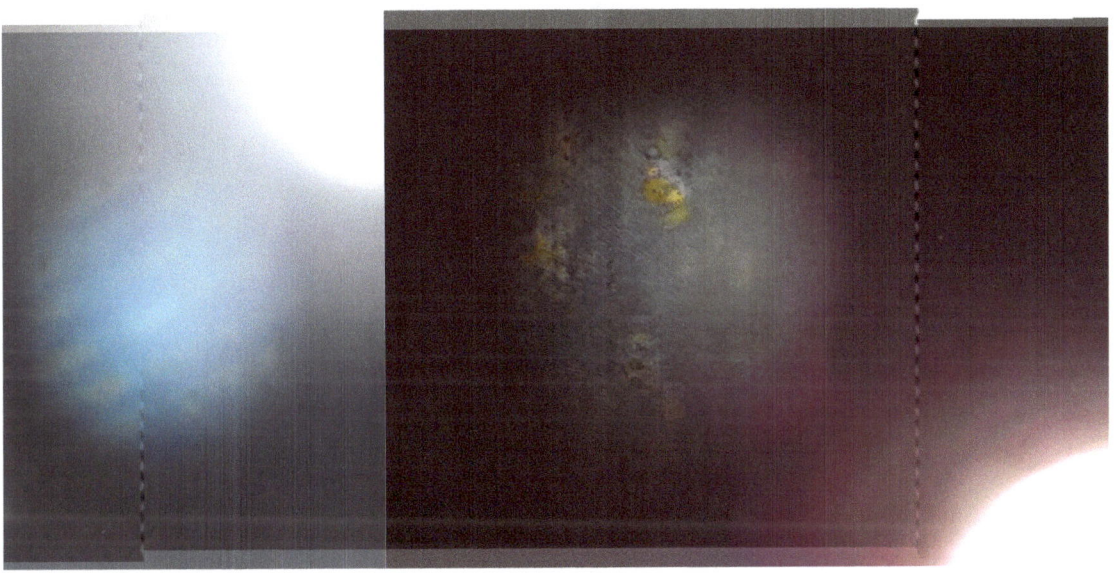

Figure 7.2: The Blue Stellar Core photographed, in the Sun's corona, through a telescope, in May 2017 (left) and in July of 2017 (right). The object shed much of the material that made up its stripes, seen in the earlier image, in the ensuing time, indicating that Stellar Cores shed material, thus creating debris in the inner Solar System. The debris is likely to be in different sizes, from planetary sized to microdust sized.

Because the objects are energy depleted they will absorb energy from any object that has more energy than they do, and the earth, since it is still able to generate energy in its core, is a constant attraction. The energy absorption process causes water in the earth's atmosphere to lose energy, condense into droplets, and thus form clouds around Stellar Core matter. In addition, the particles in the clouds continue to give off light, which is absorbed by the Stellar Core matter, but enough of this light comes from the cloud to allow us to see it as strange bright colors; pink, red, yellow, blue, orange, etc. (see Article 338: The Planet X effect: heating and ionization in contact regions) [3].

These strange colors are seen mainly at sunset but nowadays they are sometimes seen long before sunset. They are mainly seen at sunset because during the day the sun simulation system emits too much light for these colors to be clearly discernible. They are been seen at other times because more and more Stellar Core matter is arriving. They are not seen during the night because the Stellar Cores and Stellar Core debris are coming from the Sun. The dust will most likely travel with the larger objects from the Sun outwards. The fact that more is arriving may indicate that more and more of these objects are invading the Solar System. But, since these objects arrive at the Sun as comets, that would require that the number, of observed comets, would have greatly increased in the last year, or so. However, that does not seem to be the case, as the numbers have been very high, but steady, in about the last 10 years. Therefore, the great increase can only mean a huge change with the Sun itself and that can only be something to do with it weakening (see Article 372: The Sun is no longer shining: satellite images) [4]. A weaker sun is not as attractive to these objects, which will thus seek other sources of energy, such as the Earth.

In conclusion, Stellar Core debris is continuously arriving from the Sun, which is why the bright color emission, which they produce once inside the earth's atmosphere, is only seen during daylight hours. More and more Stellar Core matter seems to be coming from the Sun suggesting that the Sun has reached a new level of weakness, recently.

References:

[1]   Albers, C. (2018). Article 376: What caused the bright pink sky in Florida during the Hurricane Michael event?

[2]   Albers, C. (2018). Article 275: Planet X debris field impacting earth (in Book 8: Planet X and the Solar System).

[3]   Albers, C. (2018). Article 338: The Planet X effect: heating and ionization in contact regions ((in Book 8: Planet X and the Solar System).

[4]   Albers, C. (2018). Article 372: The Sun is no longer shining: satellite images.

Planet X Orbiting Our Sun Photo Captures by Scott C'one from Planet X News

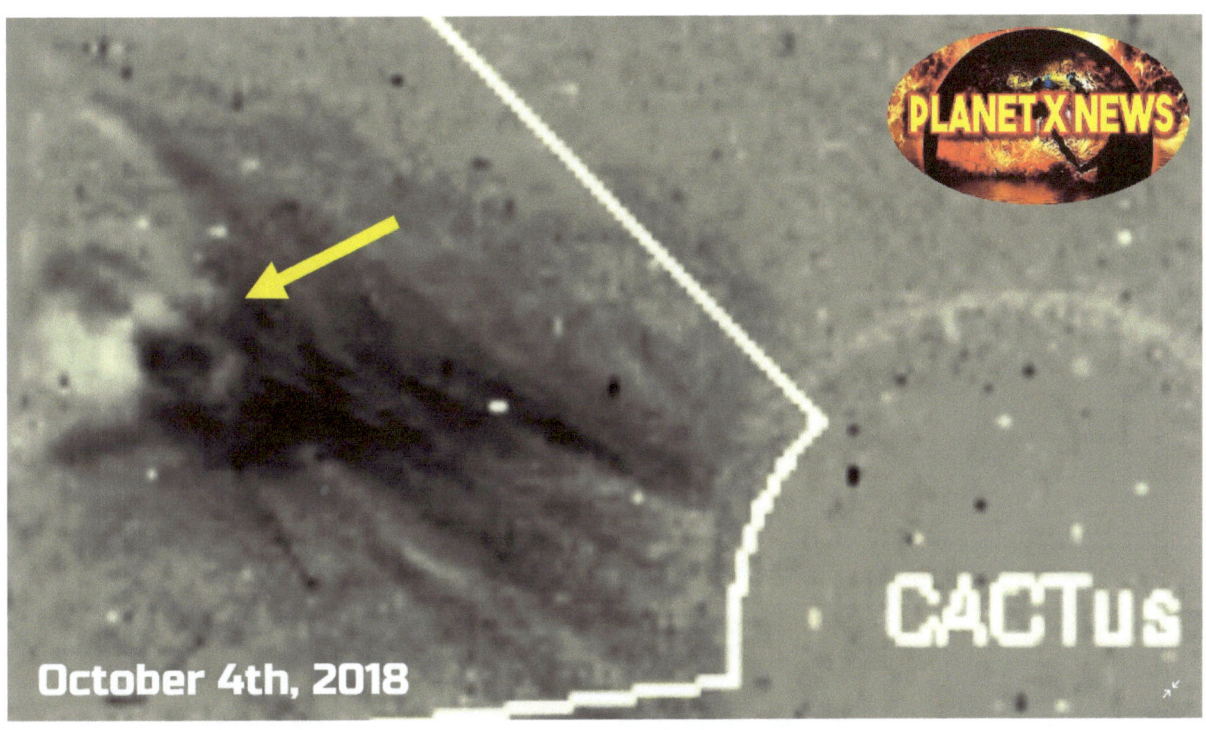

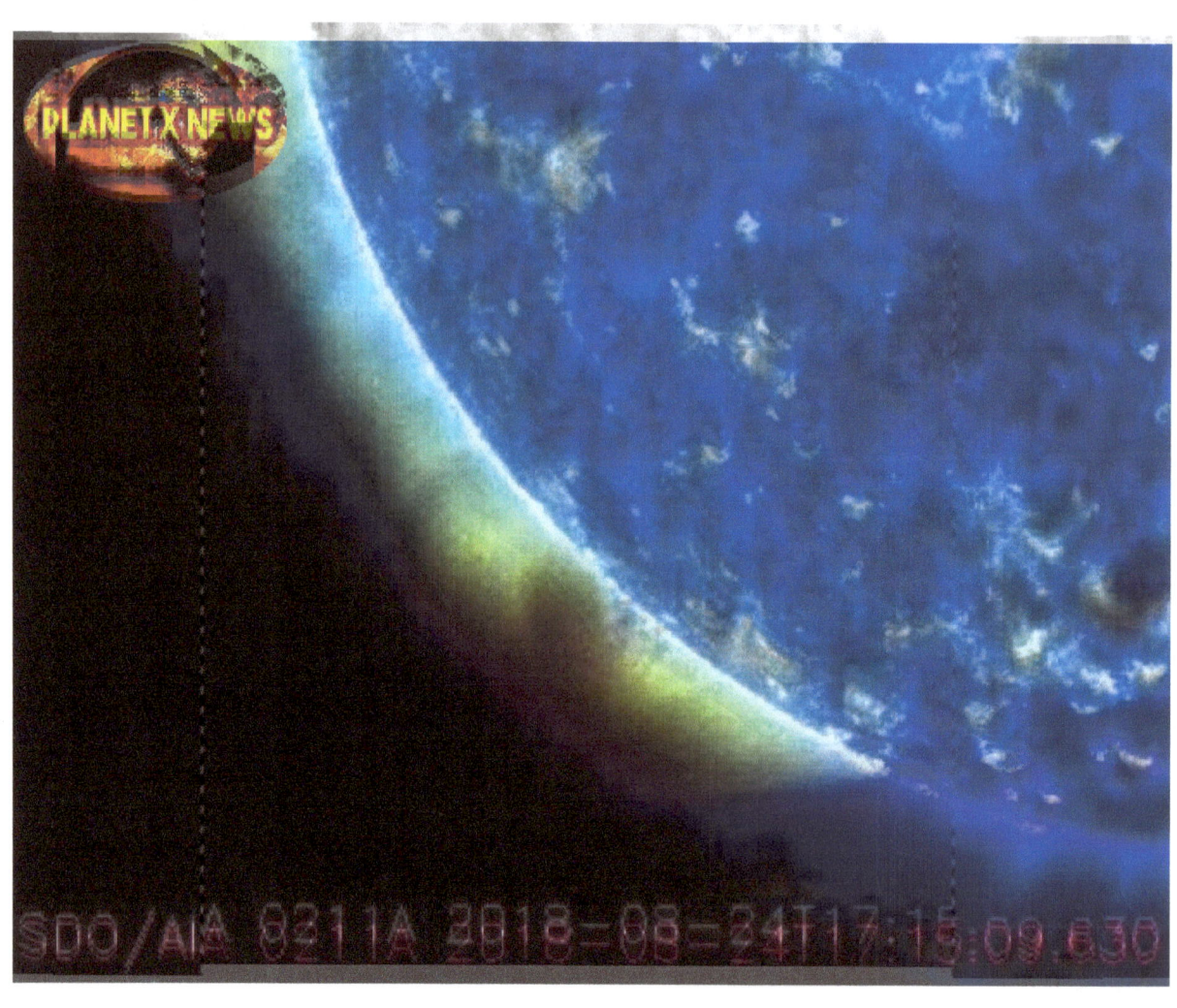

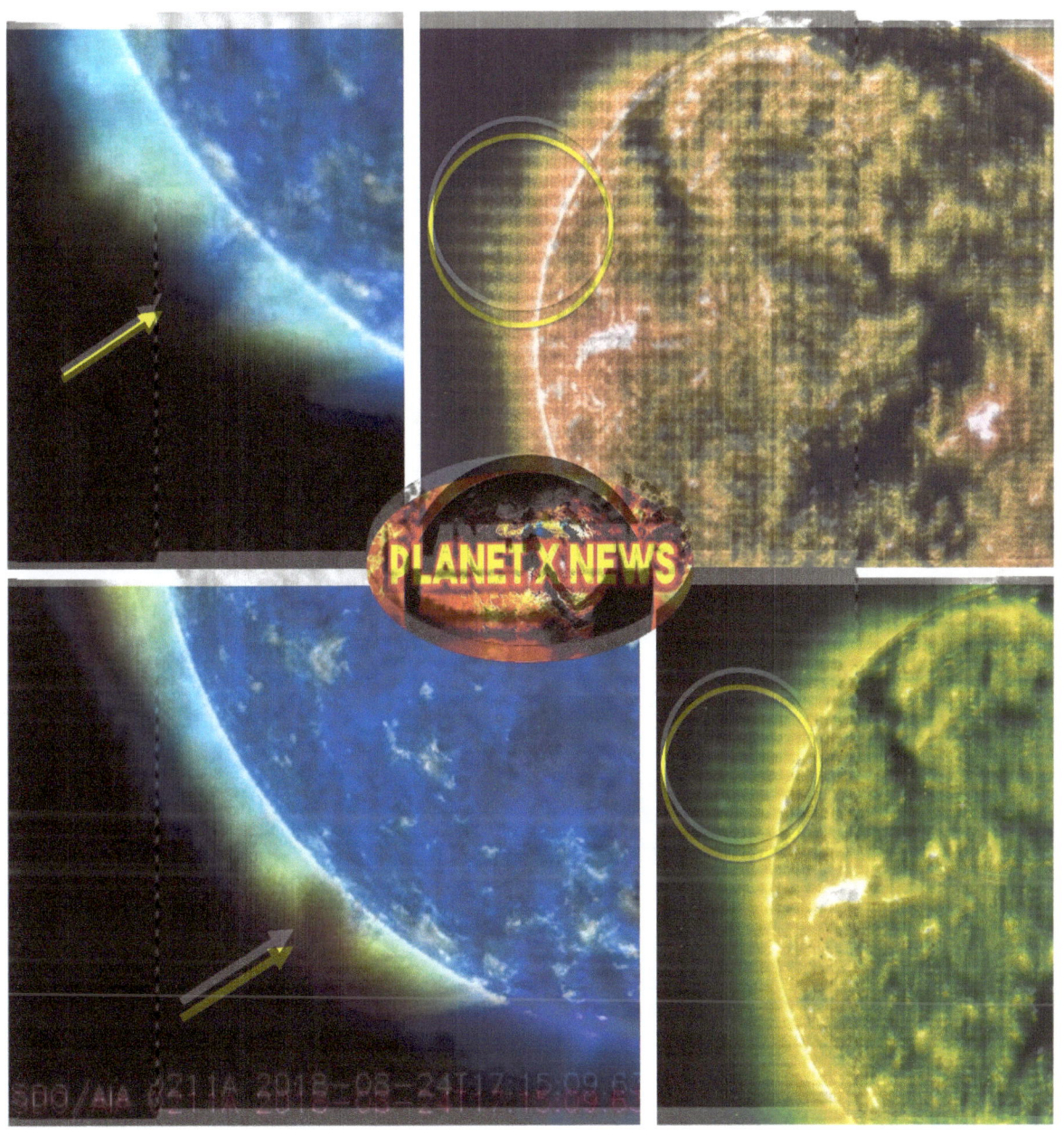

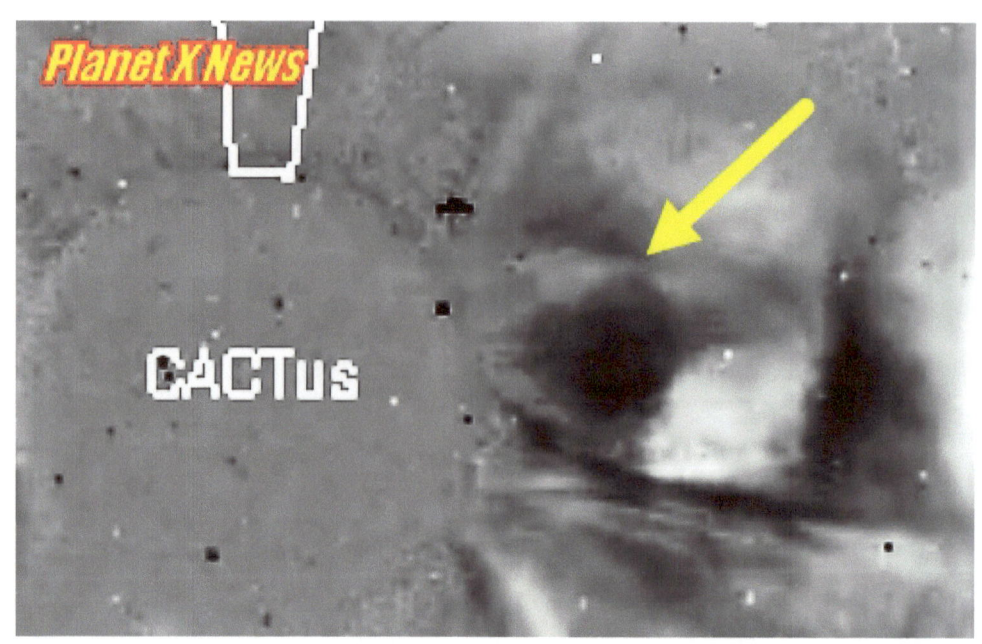

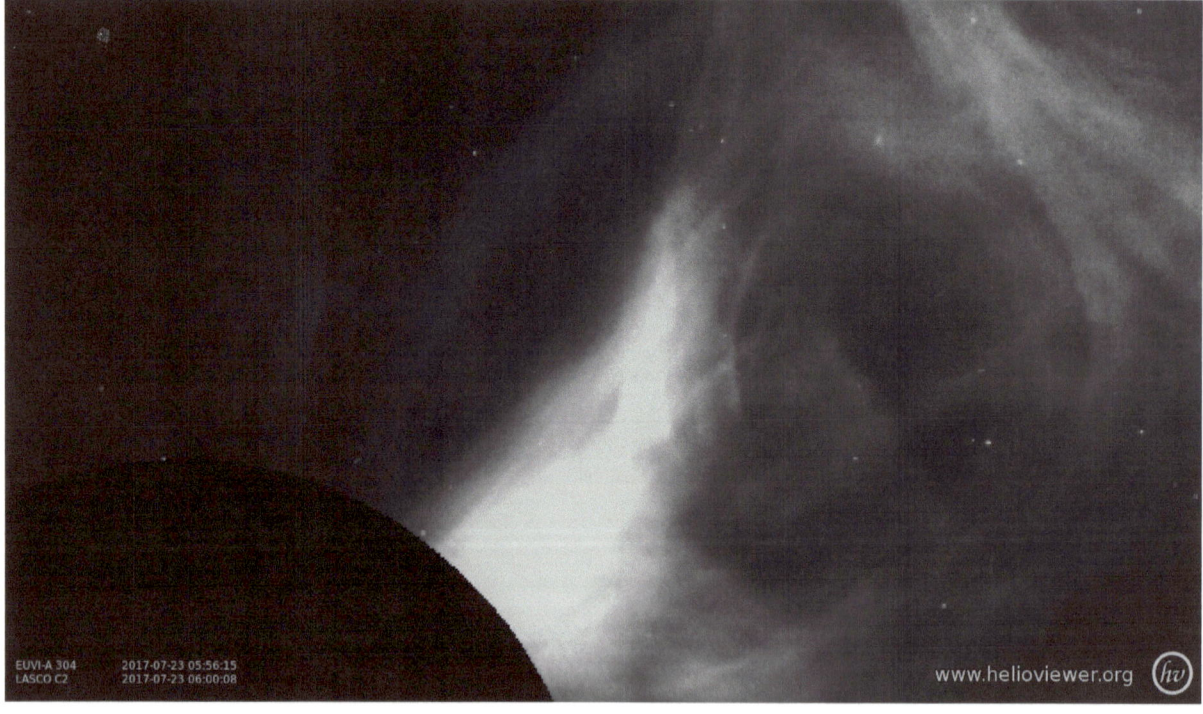

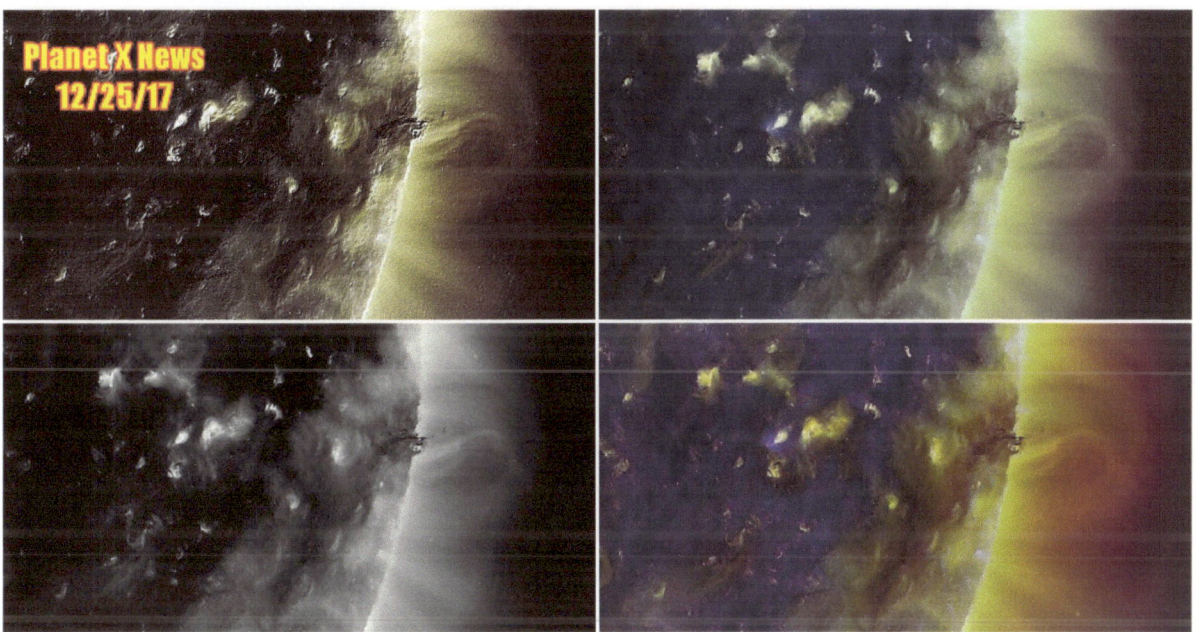

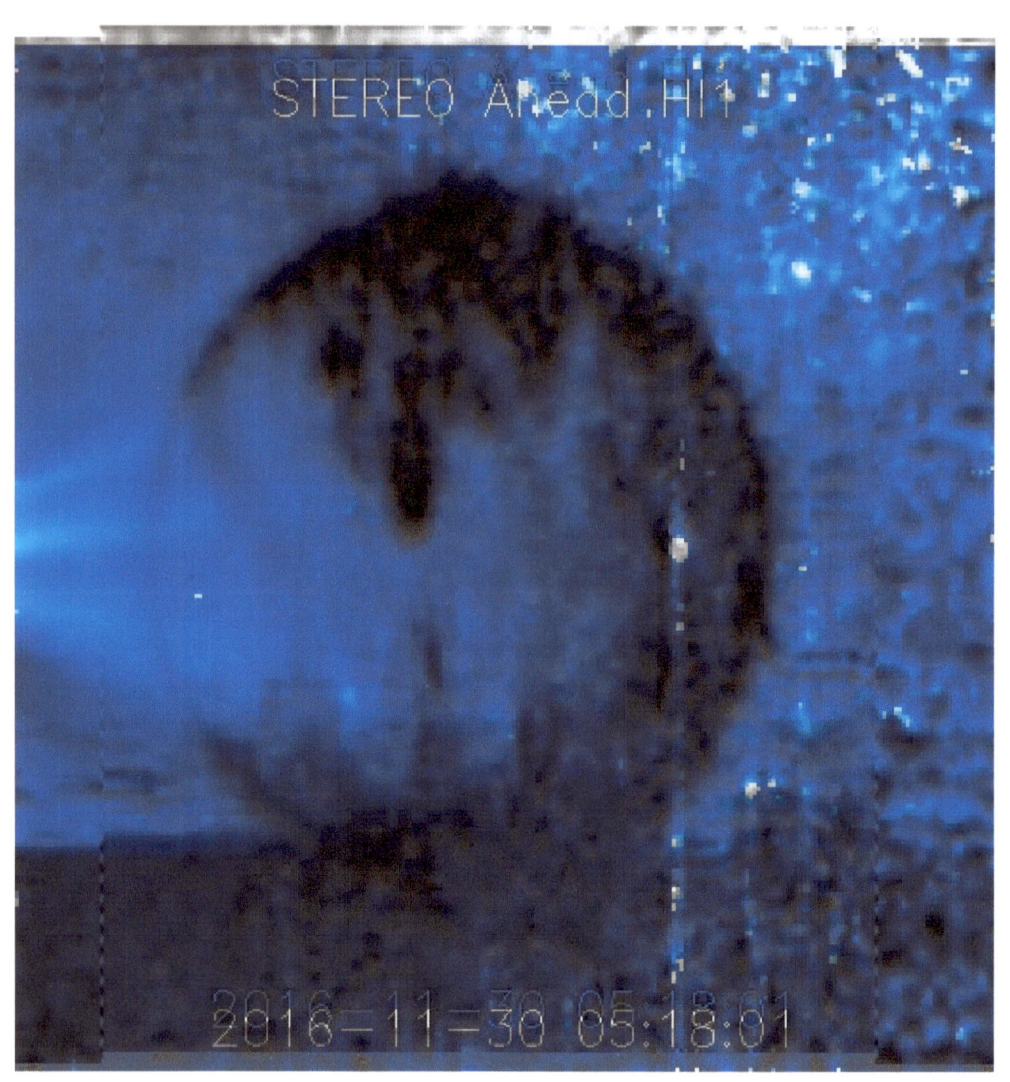

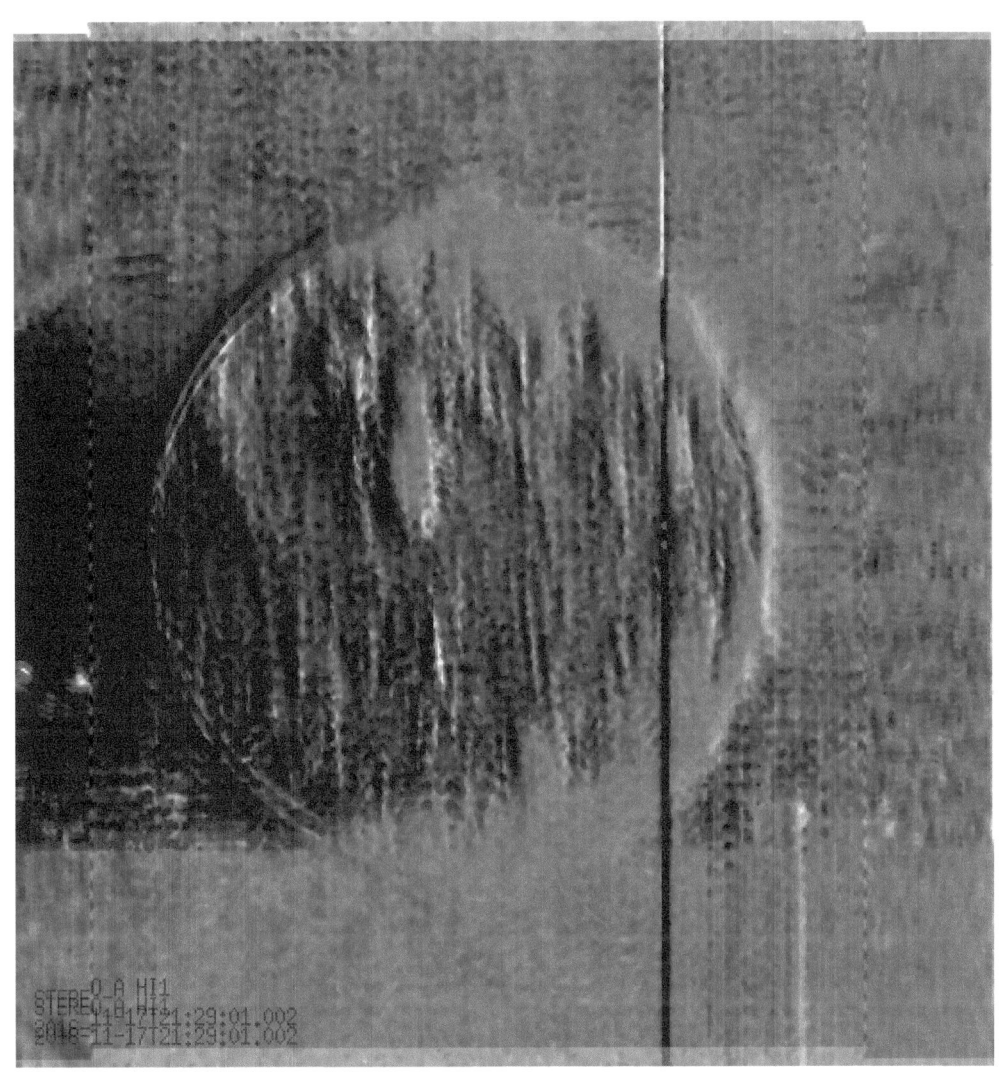

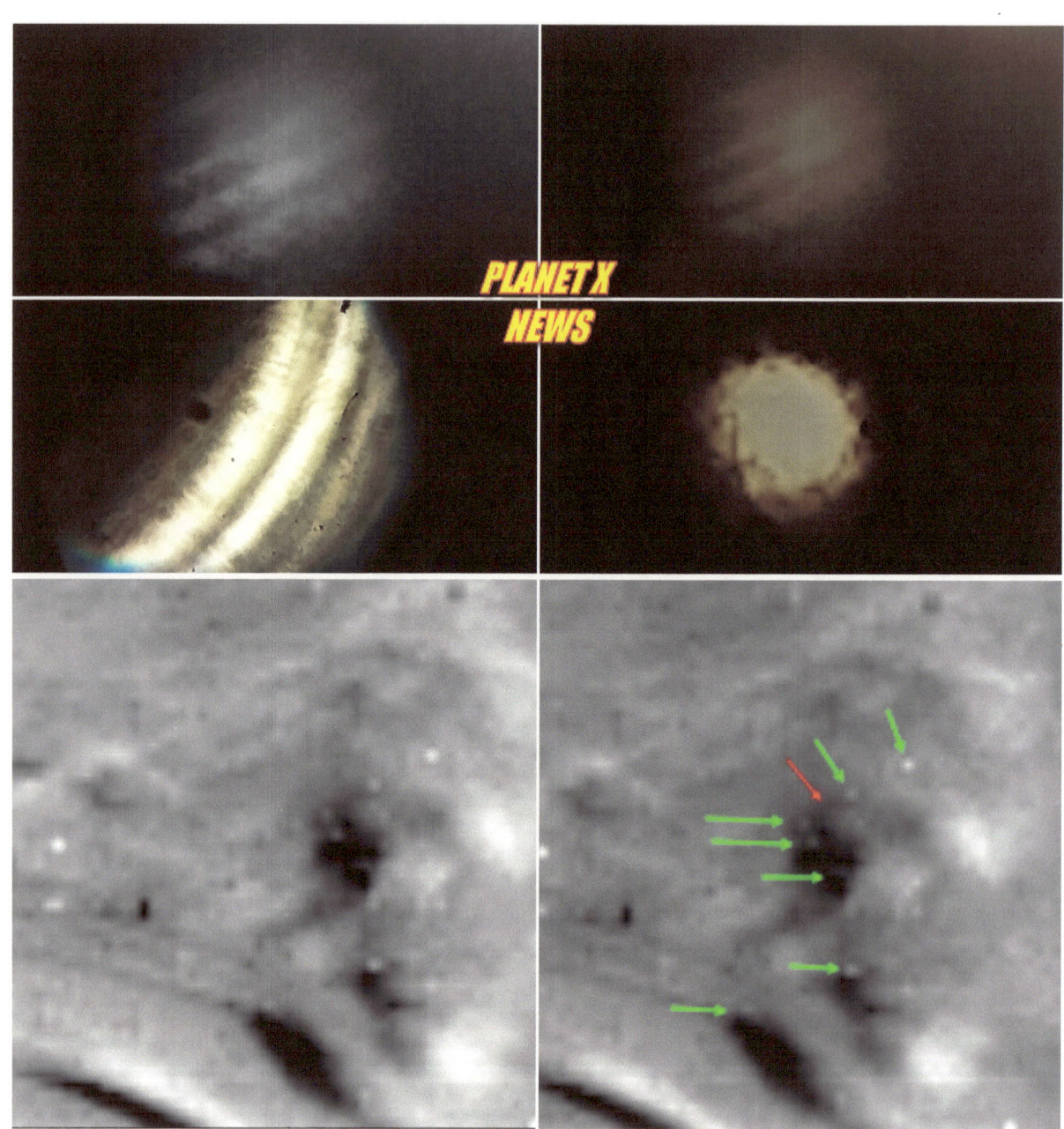

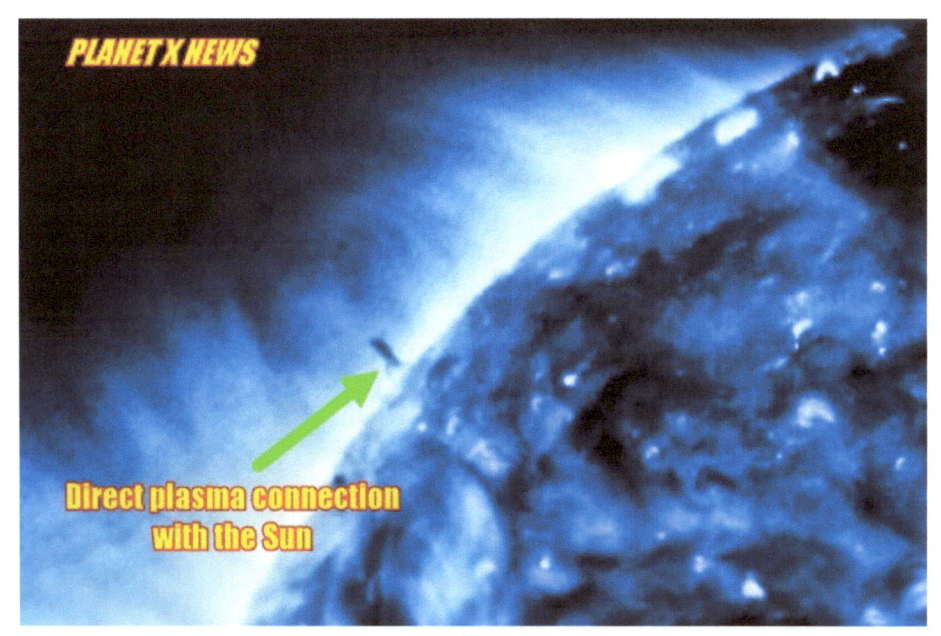

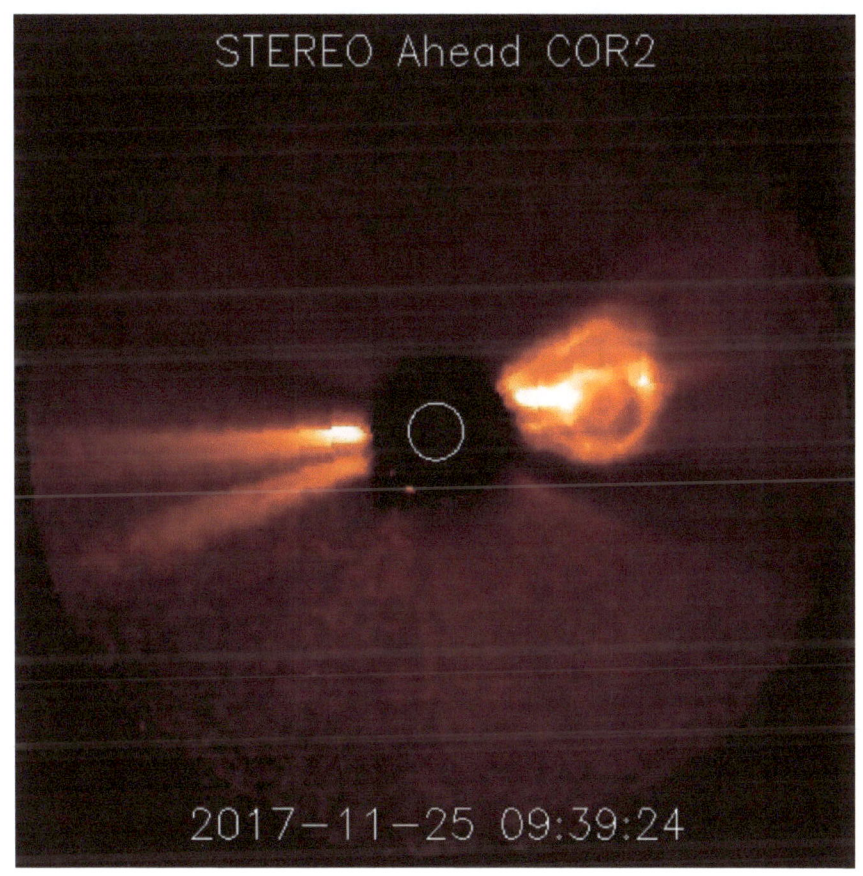

REPORTS ON HURRICANE MICHAEL AND THE DEVASTATION

Hurricane Michael was the third-most intense Atlantic hurricane to make landfall in the United States in terms of pressure, behind the 1935 Labor Day hurricane and Hurricane Camille of 1969. It was also the strongest in terms of maximum sustained wind speed to strike the contiguous United States since Andrew in 1992. In addition, it was the strongest on record in the Florida Panhandle, and was the fourth-strongest landfalling hurricane in the contiguous United States, in terms of wind speed.

The thirteenth named storm, seventh hurricane, and second major hurricane of the 2018 Atlantic hurricane season, Michael originated from a broad low-pressure area that formed in the southwestern Caribbean Sea on October 2.

The disturbance became a tropical depression on October 7, after nearly a week of slow development. By the next day, Michael had intensified into a hurricane near the western tip of Cuba as it moved northward.

The hurricane strengthened rapidly in the Gulf of Mexico, reaching major hurricane status on October 9, peaking at a Category 4 hurricane on the Saffir–Simpson scale.

Approaching the Florida Panhandle, Michael attained peak winds of 155 mph (250 km/h) as it made landfall near Mexico Beach, Florida, on October 10, becoming the first to do so in the region as a Category 4 hurricane, and making landfall as the strongest storm of the season. As it moved inland, the storm weakened and began to take a northeastward trajectory toward Chesapeake Bay, weakening to a tropical storm over Georgia, and transitioning into an extratropical cyclone off the coast of the Mid-Atlantic states on October 12.

Michael subsequently strengthened into a powerful extratropical cyclone and eventually impacted the Iberian Peninsula, before dissipating on October 16.

By October 16, at least 48 deaths had been attributed to the storm, including 33 in the United States and 15 in Central America. Insurance losses due to Michael in the United States are estimated to be at least $6 billion (2018 USD).[1] As a tropical depression, the storm caused extensive flooding in Central America in concert with a second disturbance over the eastern Pacific Ocean.

In Cuba, the hurricane's winds left over 200,000 people without power as the storm passed to the island's west. Along the Florida panhandle, the cities of Mexico Beach and Panama City suffered the worst of Michael, with catastrophic damage reported due to the extreme winds and storm surge.

Numerous homes were flattened and trees felled over a wide swath of the panhandle. A maximum wind gust of 129 mph (208 km/h) was measured at Tyndall Air Force Base near the point of landfall. As Michael tracked across the Southeastern United States, strong winds caused extensive power outages across the region.

A week after Hurricane Michael hit, more misery and a rising death toll

A week after Hurricane Michael slammed the Florida Panhandle, the scope of the storm's fury is still emerging as the death toll rises and rescuers search for the missing in the hardest-hit areas.

Michael has killed at least 32 people across Florida, Georgia, North Carolina and Virginia. Of those, 15 were in Florida's Bay County, where the hurricane made landfall last week as a Category 4 storm.

Authorities fear some people who did not evacuate could be buried beneath piles of concrete, wood and mangled metal in Florida.

The Florida Department of Health provided an online form to report those who are still unaccounted for, trapped or in need of help. While the exact number of the missing was not immediately available, more details are expected to emerge as electricity and phone services are gradually restored across the Panhandle.

Bay County, which was badly hit, has linked 15 deaths to the Category 4 storm, up from four Monday. Details on how the people died were not immediately available.

Sheriff's deputies said they're finding more victims as the waters recede.

Of the additional bodies found Tuesday, one was discovered by a K-9 in the Mexico Beach area, while a drone unit found another one.

The mountain of debris scattered in neighborhoods is complicating rescue and recovery efforts. Across Bay County, more than 2,500 structures are damaged, and at least 162 have been destroyed, authorities said.

Bay County Sheriff Tommy Ford said he does not have an accurate number of the missing people due to the lack of communications in affected areas.

Gov. Rick Scott's request for the county's transitional shelter has been approved, and federal officials will provide more options for families to stay, including hotel rooms, condos and other rentals.

Not everyone has seen what's left of their homes

In the obliterated town of Mexico Beach, some residents have visited their properties, surveyed the damage and retrieved some items. Many more are expected Wednesday.

While they aren't allowed to stay just yet, they can show the National Guard troops their IDs and visit their neighborhoods to get an idea of what it will take or if they want to rebuild.

Lisa Patrick is overcome with emotion as she sees what was left of her home to see if she can salvage anything after it was destroyed by Hurricane Michael.

Lisa Patrick is overcome with emotion as she sees what's left of her home to see if she can salvage anything after it was destroyed by Hurricane Michael.

In some neighborhoods, all that's left are concrete slabs and piles of wood, and residents have to guess the general location of where their homes once stood.

Chad Frazier has plenty of reasons to be devastated. His business was wiped out by the hurricane. His son's middle school in Panama City was also annihilated.

But Frazier said he is too thankful to be upset. Immediately after the hurricane, strangers came from out of town to provide food and water.

"The people who came, that was the biggest blessing to us," he said.

As he stood in front of a mountain of debris, Frazier said he's coping well, thanks to the outpouring of generosity.

"My shop is down, (but) I'm not in bad spirits," he said. "This just made my faith grow."

Cell phone service is trickling back on

After days of no cell reception, some residents were able to use their phones this week.

At Inlet Beach, near Panama City, excited drivers pulled over on the side of the road at the first sign of cell reception, CNN affiliate WEAR reported Tuesday. It said many of them hadn't had power and cell service for days.

In some areas, people used their cell phones for the first time when a company brought in a flying COW -- a Cell On Wings -- a drone that serves as an LTE site.

A woman journeys back to her wrecked home after Michael. This is her chronicle

A woman journeys back to her wrecked home after Michael. This is her chronicle

While service has resumed in some places, some still have no cell reception.

"Due to these outages, families are having a difficult time communicating with loved ones, first responders have faced challenges communicating and people are having difficulty getting their prescriptions filled because of the inability to connect to a network," Scott said Tuesday.

The governor demanded that telecommunications companies provide a restoration plan.

In addition to lack of cell service, more than 158,000 power customers didn't have electricity Tuesday -- nearly 140,000 of them in Florida.

Some still have no food or power

Residents in the hardest-hit areas are relying on airdropped food and water to survive.

Residents in the hardest-hit areas are relying on airdropped food and water to survive.

The Federal Emergency Management Agency has 12 teams in Florida to help people register for disaster assistance. It also has 42 distribution points throughout Florida and Georgia where people can get food and water.

About 1,200 people are in shelters, state officials said. Drivers have been lining up for hours to get fuel.

Some school districts are still closed

The Florida Department of Education said many schools in affected areas don't have power, and authorities are working to ensure it's restored quickly. As a result, several school districts are closed until further notice, including Calhoun and Washington counties.

Bay County hopes to open in mid-November but, with some buildings too badly damaged, other facilities will have split days with two half-day sessions as regular students share schools with displaced pupils. Some students will have to take long bus rides to schools in Panama City, said Steve Moss, vice chairman of the Bay County School District.

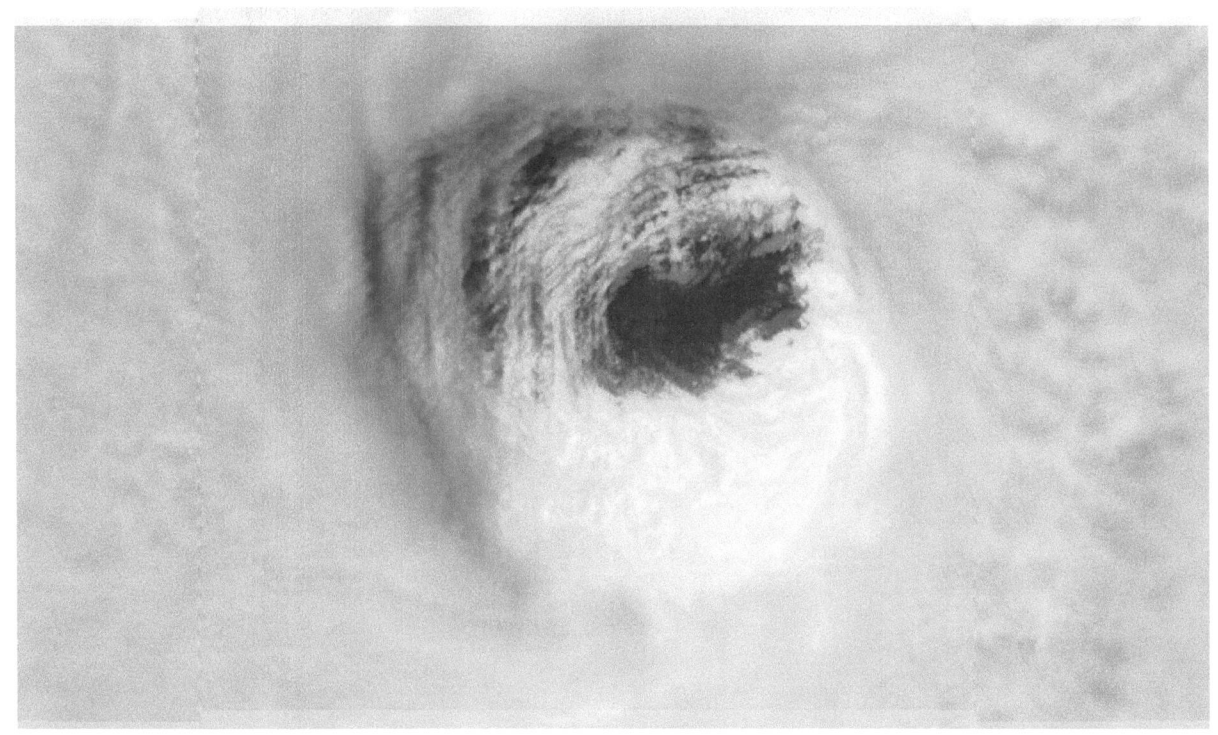

Florida's Panhandle is covered with thousands of pine trees. And as winds top 150 mph, those trees could turn into violent projectiles.

"You get those kinds of winds, (and) it's catastrophic damage to the trees' structure," Graham said.

And with downed trees come power outages.

"This could bring down thousands and thousands of those pine trees here -- not only making all the damage along the coast but inland as well," CNN meteorologist Chad Myers said.

Graham said because this storm is "absolutely overwhelming," power outages could last weeks.

Cities far inland will feel an actual hurricane

Many hurricanes sputter out after they hit land and lose the title of hurricane.

Michael plowed through the Southeast as a hurricane, with winds topping 73 mph as it crossed into Georgia. By early Thursday, it was a tropical storm with winds of 70 mph

"Because of the forward movement -- the decent forward movement it has -- you're going to see a hurricane stay intact through southwest and central Georgia," Long said.

"And then you're going to see rainfall through South and North Carolina, dumping 4 to 6 inches of rain in rivers that are already saturated and haven't really receded much from Florence a few weeks ago."

Storm surge was deadly

Michael will spawn massive storm surges -- or walls of ocean water -- as high as 14 feet, forecasters said.

"Half of the fatalities in these tropical systems occur with the storm surge," Graham said.

The FEMA chief said storm surge is huge reason why evacuations are so critical.

"This is nothing to play around with," Long said. "Those who stick around and experience storm surge are less likely to live to tell about it."

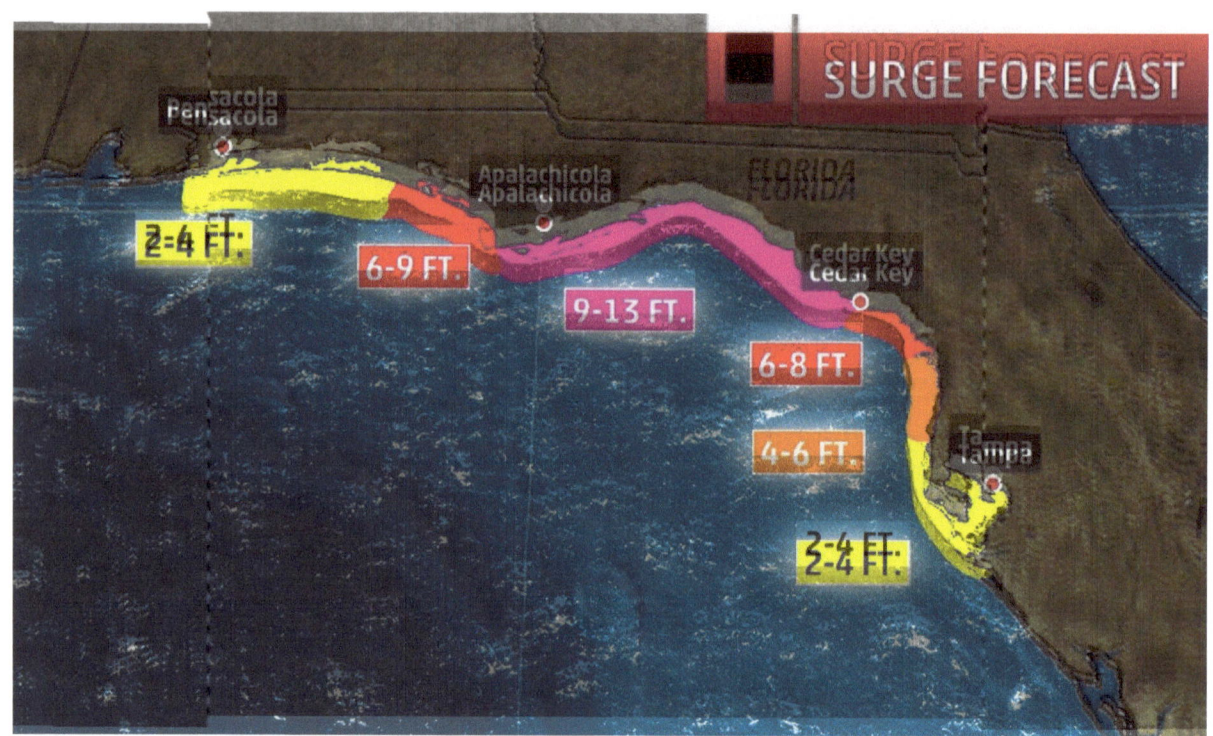

AFTER GATHERING STRENGTH from the warm waters of the Gulf of Mexico overnight, Hurricane Michael blasted across the Florida Panhandle Wednesday afternoon, pummeling the area with winds up to 155 miles per hour. That makes the Category 4 hurricane one of the all-time strongest landfalls in US history.

Earlier today, NOAA's National Hurricane Center warned that in addition to the destructive winds and heavy rains, Michael could bring a storm surge of up to 14 feet to areas in the direct path of the storm. Surge is similar in effect to a tsunami—a wall of water created when atmospheric pressure changes cause the ocean to rapidly rise and high winds push all that water onshore. It's measured as the height of the water above the normal predicted tide, and how bad it is depends mainly on three things: wind speed, shoreline shape, and timing.

A destructive and life-threatening storm surge event will occur along portions of the Florida Panhandle, Big Bend, and Nature Coast on Wednesday and Wed night. The worst storm surge is expected to be between Mexico Beach and Keaton Beach where 9-13' of inundation is possible.

Typically, the strongest surge occurs with the eyewall of the storm. That makes sense, the strongest wind is where you see the strongest shove. Michael made landfall at Mexico Beach, just northwest of a jutting headland and the town of Apalachicola. But the topography of the area, and of the underwater terrain offshore, slowed down the water's rise. A National Ocean Service water level station at Apalachicola recently reported about 6.5 feet of inundation.

That's still enough to knock you off your feet or send cars and other large objects hurtling headlong into whatever is in the water's path.

And it's why storm surge is often the deadliest aspect of hurricanes. As you can see in this Weather Channel simulation, courtesy of Climate Desk, surviving surges more than nine feet is unlikely.

To the east of Apalachicola, the winds were weaker. But the concave shape of the coastline in that part of the state threatened to push the water to dangerous heights. "It gets pinched tighter and tighter and higher and higher as the water goes over an increasingly narrow space," says Jamie Rhone, the storm surge specialist at the National Hurricane Center in Miami, Florida, in charge of the agency's predictions. "That's why we saw the 9-to-14 foot forecast extend so far east of the eyewall."

During the hurricane off-season, Rhone and other NOAA meteorologists build the models that they use to make such forecasts. That means updating them to reflect the current morphological state of the coasts—shifting sand dunes, eroding beaches, human development.

All of those things impact the path water can take during a big storm. The models also need to take into account the location of the ocean itself. Since 1970, Apalachicola sea levels have risen about a third of a foot. While the NHC doesn't produce estimates of sea level rise, its models take into account the current conditions.

"Surge rides on top of the preexisting ocean, so if it's any higher now than it was 50 years ago, the surge is going to be that much greater," says Rhone.

The third factor is when the storm hits. The higher the tide, the more dangerous the storm surge gets. Luckily for Florida residents, Michael made landfall during low tide. It could have been much worse.

As the storm recedes from the coast and heads inland, a surge warning is still in effect for about 400 miles of coastline, from the Okaloosa/Walton County Line to the Anclote River.

Rhone's team is working on surge forecasts further north, as Michael makes its way through Georgia and the Carolinas.

At the same time, they're taking a first crack at modeling how bad the damage is in Florida, to give first responders a sense of what to expect as they head into the worst-hit areas.

Hurricane Michael's Storm Surge Eroded 75 Percent of Florida Panhandle Beaches, USGS Says

USGS Report

Hurricane Michael is expected to make landfall Wednesday in the Florida Panhandle as a Category 3 storm and is very likely to cause erosion at the base of sand dunes of three-quarters of the beaches along the panhandle and flood more than a quarter of its dunes, causing flooding behind protective dune lines, according to the U.S. Geological Survey.

"As the storm approaches the coast, the shallow seabed will reduce that wave height," said Doran. "But water levels in some parts of the Northeast Florida coast will still be high enough to overwhelm the dunes, which are relatively low and narrow."

The area from Cape San Blas to St. George is of particular concern, Doran added. The area largely consists of beach communities and state parks, which are forecast to have a total water level 16 to 20 feet higher than normal. If Michael's forecast proves right, there's a 95 percent chance the area's dunes will be flooded by storm waters.

Overall, it's highly likely that Michael's storm surge will cause erosion at the base of the dune line for 75 percent of Florida Panhandle beaches.

Of the threats a major storm can pose, beach erosion is just the first level of damage.

Forty-three percent of Florida Panhandle beaches and 17 percent of Pinellas and Manatee County beaches are forecast to encounter overwash, which occurs when waves and surge top dunes and transport large amounts of sand across coastal environments, significantly changing the landscape.

The most severe impact a storm can pose is inundation, when beaches and dunes are completely submerged by surge for an extended period of time. Twenty-seven percent of Florida Panhandle beaches and 17 percent of Pinellas and Manatee counties are forecast as very likely to see inundation.

FRANKLIN COUNTY, Fla. (WCTV) -- Piles of oyster shells, wood, patio furniture and even a refrigerator sit on the side of Highway 98 in Franklin County.

The debris was swept from Alligator Point across the Ochlockonee Bay and across the highway to the inland side of the road.

The National Weather Service in Tallahassee noted where the debris moved, while surveying damage. The group in Franklin County is one of four teams surveying damage along the coast and inland through Southwest Georgia.

Kelly Godsey is a meteorologist and service hydrologist for NWS Tallahassee. He and his team are primarily looking for high water lines. They use a laser level to measure the debris or water line to sea level.

"Down at the coast, we're seeing numbers that are very similar to what we were forecasting, in the 9-to-12 foot range," said Godsey. "We talk about above ground level right on the waterfront. As we move inland, some of the preliminary numbers that we've seen have been around 4 to 7 feet above ground."

ST MARKS — Despite the sunny sky, the trek to the storm tide sensors at the San Marcos State Park was more than just muddy. U.S. Geological Survey intern Wyatt Timmons described it as "mucky."

USGS scientists deployed more than 30 storm tide sensors throughout the Florida Gulf Coast before Hurricane Michael made landfall, and their first stop after the storm was San Marcos Public Park Boat Ramp.

Three USGS crew members took on the challenge of getting to the base of a damaged boating dock where a storm tide sensor awaited. Stumbling through the crab infested debris with mud to their ankles, the crew made it to the sensor, a device that can determine the depth, duration and timing of the storm.

"**Storm tide sensors** measure water level data and barometric pressure at 30-second intervals," said hydrologic technician Lori Lewis.

Although the storm-tide sensors can digitally record the water levels, Lewis and her team believe in double checking by looking for high water lines created by foam, seeds or other debris.

"This can be anywhere where it's being protected, where the waves haven't washed them away," she said, noting they found a seed line in the San Marcos State Park restroom.

"Sometimes we find them in buildings, like we did today, on the backs of telephone poles or signs. We're looking for a definitive line that will mark and determine the elevation."

Death as record-breaking Hurricane Michael slams into Florida

The most powerful hurricane on record to hit Florida's Panhandle has left wide destruction and at least one person dead as it crossed Georgia toward the Carolinas.

Authorities said a man was killed by a tree falling on a Panhandle home, while search and rescue crews were expected to escalate efforts to reach hardest-hit areas and check for anyone trapped or injured in the storm debris.

A day after the supercharged storm crashed ashore amid white sand beaches, fishing towns and military bases, Michael was no longer a Category 4 monster packing 155 mph winds.

Downgraded to a tropical storm early on Thursday over south-central Georgia, it continued to weaken but was still menacing the Southeast with heavy rains, blustery winds and possible spin-off tornadoes.

After daylight on Thursday residents of north Florida would just be beginning to take stock of the enormity of the disaster.

Damage in Panama City near where Michael came ashore on Wednesday afternoon was so extensive that broken and uprooted trees and downed power lines lay nearly everywhere.

Roofs were peeled away, sent airborne, and homes were split open by fallen trees, while more than 380,000 homes and businesses were without power at the height of the storm.

Vance Beu, 29, was staying with his mother at her home, Spring Gate Apartments, a complex of single-story wood frame buildings where they piled up mattresses around themselves for protection.

A pine tree punched a hole in their roof and his ears even popped when the barometric pressure went lower. The roar of the winds, he said, sounded like a jet engine.

"It was terrifying, honestly. There was a lot of noise. We thought the windows were going to break at any time," Mr Beu said.

Sally Crown rode out Michael on the Florida Panhandle thinking at first that the worst damage was the many trees downed in her yard. But after the storm passed, she emerged to check on the cafe she manages and discovered a scene of breathtaking destruction.

"It's absolutely horrendous. Catastrophic," Ms Crown said. "There's flooding. Boats on the highway. A house on the highway. Houses that have been there forever are just shattered."

Governor Rick Scott announced that thousands of law enforcement officers, utility crews and search and rescue teams would now go into recovery mode.

"Hurricane Michael cannot break Florida," Mr Scott vowed.

Michael sprang quickly from a weekend tropical depression, going from a Category 2 on Tuesday to a Category 4 by the time it came ashore.

It forced more than 375,000 people up and down the Gulf Coast to evacuate as it gained strength quickly while crossing the eastern Gulf of Mexico toward north Florida.

Based on its internal barometric pressure, Michael was the third most powerful hurricane to hit the US mainland, behind the unnamed Labor Day storm of 1935 and Camille in 1969.

Based on wind speed, it was the fourth-strongest, behind the Labor Day storm (184 mph), Camille and Andrew in 1992.

It also brought the dangers of a life-threatening storm surge.

In Mexico Beach, with a population of 1,000, the storm shattered homes, leaving floating piles of lumber.

"We are in new territory," National Hurricane Centre Meteorologist Dennis Feltgen wrote on Facebook.

"The historical record, going back to 1851, finds no Category 4 hurricane ever hitting the Florida panhandle."

After Michael left the Panhandle late on Wednesday, Kaylee O'Brien was crying as she sorted through the remains of the apartment she shared with three roommates at Whispering Pines apartments.

Four pine trees had crashed through the roof of her apartment, nearly hitting two people, and her one-year-old Siamese cat, Molly, was missing.

"We haven't seen her since the tree hit the den. She's my baby," Ms O'Brien said.

Hurricane Michael made landfall in the Florida Panhandle Wednesday afternoon. The intense Category 4 hurricane was packing top sustained winds of 155 mph when it crashed ashore in the early afternoon near Mexico Beach.

The National Hurricane Center described Michael as "potentially catastrophic." Michael was the worst storm ever to hit the Panhandle.

Nearly 30 million people in the Southeast were in its crosshairs. Forecasters said Michael was bringing damaging winds and potentially life-threatening storm surge.

Hurricane Michael by the numbers

- Hurricane history: first Category 4 hurricane to make landfall in Florida's Panhandle since record-keeping began in 1851. With a minimum pressure of 919 millibars in the hurricane's eye, it was the third most intense hurricane landfall in the U.S. in recorded history
- Wind speeds at 7 p.m.: 100 mph, with gusts topping 60 mph at several Georgia airports
- Current location: 35 miles west-southwest of Albany, Georgia
- High tides: storm surge of 6 feet up to 14 feet forecast for Florida's Panhandle and Big Bend
- Get out: roughly 375,000 people in Florida warned to evacuate
- Staying safe: nearly 6,700 people took refuge in 54 shelters in Florida
- Power outages: 370,060 customers without power in Florida, Alabama and Georgia
- Food and water: 1.5 million ready-to-eat meals, 1 million gallons of water and 40,000 10-pound bags of ice ready for distribution

'I don't have anything left': Hurricane Michael survivors scramble for food, water as death toll rises

PANAMA CITY, Fla. - While crews continued the search Saturday for thousands of people reported missing after Hurricane Michael ravaged Florida's Panhandle, those devastated by the storm were left scrambling for food and water – trying to put the pieces of their lives back together.

The death toll from the monstrous storm has risen almost daily as crews made their way into some of the areas hit hardest. As of Saturday evening, it had risen to 17.

Virginia State Police said in a news release Saturday evening that a woman's body was found earlier in the day. The state Department of Emergency Management says the discovery brings the total of storm-related deaths in Virginia to six.

By some estimates, nearly 300 people stayed behind, ignoring evacuation pleas, as the storm rolled ashore as a monster Category 4. Officials have gotten thousands of missing people reports but with network outages, it's been a tough job determining whether some of the missing are simply unable to communicate with loved ones.

Long lines have formed at distribution centers where authorities are giving out food and water to those in need. The National Guard helped Saturday to man a station at Lucille Moore Elementary School in Panama City, Florida.

The troops, out of Sanford, Florida set up about a dozen pallets of bottled water early Saturday. But more than delivering supplies, the group of about 97 soldiers was trying to deliver hope.

Grateful families smiled, waved and thanked the soldiers as they loaded them up with supplies, and soldiers smiled, waved back and offered words of encouragement.

"Smiles are contagious," 2nd Lt. Scott Mandelberg explained.

Search under way for victims of deadly Hurricane Michael in US

Search and rescue teams are combing through shattered US communities looking for victims of Hurricane Michael, which carved out a swathe of destruction in the Florida Panhandle, killing at least six in three states.

In Mexico Beach, a seafront town where the hurricane made landfall, houses had been destroyed by the storm surge, boats had been tossed into gardens and the streets were littered with trees and power lines.

Florida Governor Rick Scott said the storm had caused "unbelievable devastation" and the priority for the moment was looking for survivors among residents who failed to heed orders to evacuate.

"I'm very concerned about our citizens that didn't evacuate and I just hope that, you know, we don't have much loss of life," Mr Scott said.

The US Army said more than 2,000 Florida National Guard soldiers were working on the recovery operations.

There have been six confirmed storm-related deaths so far - four in Florida's Gadsden County, one in Georgia, and one in North Carolina.

US President Donald Trump has pledged to help storm victims.

Officials said more than 400,000 homes and businesses were without electricity in Florida and Governor Scott said nearly 20,000 utility workers had been deployed to restore power.

Hurricane Michael made landfall on Wednesday afternoon as a Category 4 storm, the most powerful to hit Florida's northwestern Panhandle in more than a century.

It has since been downgraded to a tropical storm as it moves through the Carolinas, which are still recovering from last month's Hurricane Florence.

Mexico Beach suffered massive destruction from the 250km/h winds and several meters of storm surge.

Home after home was lifted from its foundations in the town of around 1,000 people, leaving just bare concrete slabs. Others were missing roofs or walls. Roads were impassable and canals were choked with debris.

Federal Emergency Management Agency chief Brock Long said Michael was the most intense hurricane to strike the Florida Panhandle since record keeping began in 1851.

Mr Long said many Florida buildings were not built to withstand a storm above the strength of a Category 3 hurricane on the five-level Saffir-Simpson Hurricane Wind Scale.

The storm posted historic numbers when it slammed into Florida about 1:30 p.m. EDT yesterday as a Category 4 hurricane. The National Hurricane Center reported Michael's sustained winds were near 155 mph as the eye of the storm moved ashore near Mexico City, FL. According to NOAA's historical hurricane tracks database, Michael is the first Category 4 storm to make landfall along the Florida Panhandle since records began in 1851.

In addition to the devastating damage that is immediately visible today, less obvious hazards in the wake of the massive storm are expected to last weeks. Food safety dangers come in various forms and can cause severe illnesses and deaths as floodwaters recede.

Among the most vulnerable foods are fresh fruits and vegetables. They are breeding grounds for pathogens when power outages cause the loss of refrigeration and temperature control. Fresh produce that comes into contact with floodwater can be instantly contaminated with a wide range of bacteria, viruses and parasites.

The toxic composition of floodwater is such a serious food safety hazard that federal law prohibits the sale, distribution or donation of any produce or other food crops from fields that are flooded. Special inspections are required before such crops can even be used for animal feed.

The Food and Drug Administration, U.S. Department of Agriculture, Centers for Disease Control and Prevention, and state and local health agencies all urge the public to destroy any home grown food that has been touched by floodwater. People in flooded areas should also exercise extreme caution when buying fresh produce from roadside stands and farmers markets in flooded areas.

Regardless if food is from a backyard garden or a large commercial farm, there is no way for it to be cleaned for safe consumption if it has been compromised by floodwater.

Other storm food safety tips from public agencies include steps to take before and after severe weather.

Steps to follow in advance of losing power:

- Keep appliance thermometers in both the refrigerator and the freezer to ensure temperatures remain food safe during a power outage. Safe temperatures are 40°F or lower in the refrigerator, 0°F or lower in the freezer.
- Freeze water in one-quart plastic storage bags or small containers prior to a storm. These containers are small enough to fit around the food in the refrigerator and freezer to help keep food cold. Remember, water expands when it freezes, so don't overfill the containers.
- Freeze refrigerated items, such as leftovers, milk and fresh meat and poultry that you may not need immediately—this helps keep them at a safe temperature longer.
- Know where you can get dry ice or block ice.
- Have coolers on hand to keep refrigerator food cold if the power will be out for more than four hours.
- Group foods together in the freezer—this 'igloo' effect helps the food stay cold longer.
- Keep a few days' worth of ready-to-eat foods that do not require cooking or cooling.

Steps to follow if the power goes out:

- Keep the refrigerator and freezer doors closed as much as possible. A refrigerator will keep food cold for about four hours if the door is kept closed. A full freezer will hold its temperature for about 48 hours (24 hours if half-full).
- Place meat and poultry to one side of the freezer or on a tray to prevent cross contamination of thawing juices.
- Use dry or block ice to keep the refrigerator as cold as possible during an extended power outage. Fifty pounds of dry ice should keep a fully-stocked 18-cubic-feet freezer cold for two days.

Food safety after a flood:

- Do not eat any food that may have come into contact with flood water—this would include raw fruits and vegetables, cartons of milk or eggs.
- Discard any food that is not in a waterproof container if there is any chance that it has come into contact with flood water. Food containers that are not waterproof include those packaged in plastic wrap or cardboard, or those with screw-caps, snap lids, pull tops and crimped caps. Flood waters can enter into any of these containers and contaminate the food inside. Also, discard cardboard juice/milk/baby formula boxes and home-canned foods if they have come in contact with flood water, because they cannot be effectively cleaned and sanitized.
- Inspect canned foods and discard any food in damaged cans. Can damage is shown by swelling, leakage, punctures, holes, fractures, extensive deep rusting or crushing/denting severe enough to

Steps to follow after a weather emergency:

- Check the temperature inside of your refrigerator and freezer. Discard any perishable food (such as meat, poultry, seafood, eggs or leftovers) that has been above 40°F for two hours or more.
- Check each item separately. Throw out any food that has an unusual odor, color or texture or feels warm to the touch.
- Check frozen food for ice crystals. The food in your freezer that partially or completely thawed may be safely refrozen if it still contains ice crystals or is 40°F or below.
- Never taste a food to decide if it's safe.
- When in doubt, throw it out.
- The USDA's Food Safety and Inspection Service has a YouTube video "Food Safety During Power Outages" that includes instructions for keeping frozen and refrigerated food safe. The publication "A Consumer's Guide to Food Safety: Severe Storms and Hurricanes" can be downloaded and printed for reference during a power outage. Infographics on FSIS' Flickr page outline steps you can take before, during and after severe weather, power outages and flooding.

After Hurricane Michael, A Call For Stricter Building Codes In Florida's Panhandle

In Mexico Beach, Fla., Lance Erwin is one of the lucky ones. His house is still standing. He stayed in his home during Hurricane Michael, several blocks from the beach, in a part of his house that he calls his "safe room."

"The garage door was shaking," he says. "I knew the roof was gone at that point because everything was shaking. I thought, 'Just hang in there.' I had faith everything was going to be OK."

Florida has some of the nation's toughest building codes. But in the Panhandle, you wouldn't know it. The rules are looser there — allowing construction that couldn't stand up to Michael's 155 mph winds.

Erwin lost doors, windows, and part of his roof. But his house is still there. The same can't be said for many other residents. Entire blocks are flattened. Mexico Beach's mayor says 75 percent of the town is gone. The storm surge washed houses on the beach off their foundations to the other side of the coastal road. Even more damaging were the high winds that lifted off roofs.

As we progress in time and our climate continues to change at a very rapid pace, storms like Hurricane Michael will become stronger and more powerful. The time is now to prepare yourself and your family for what is coming...The Awakening is Now!

www.ingramcontent.com/pod-product-compliance
Lightning Source LLC
Chambersburg PA
CBHW051914210526
45473CB00006B/2006